To future generations of scientists, technologists, engineers, mathematicians, thinkers, innovators, leaders and explorers who dare to dream.

Editorial Board

Canadian Space Agency: Marilyn Steinberg, Jason Clement
European Space Agency: Nigel Savage
Japan Aerospace Exploration Agency: Masato Koyama, Yayoi Miyagawa
Russian Federal Space Agency: Sergey Avdeev
National Aeronautics and Space Administration: Regina Blue, Judy Carrodeguas, Carolyn Knowles

Managing Editors:
Camille W. Alleyne, NASA; Susan I. Mayo, ESCG

Executive Editor:
Julie A. Robinson, NASA

Full Page Image Credits

International Space Station Education Opportunities

Student-Developed Investigations. Students from Lowery Intermediate in Louisiana study a map to compare "Delta's around the world to our Delta." *Image courtesy of EarthKAM.*

Education Competitions. Bending water. *Image courtesy of NASA.*

Students Performing Classroom Versions of ISS Investigations. View of *Nephila clavipes* (golden orb spider) inside the spider habitat. *Image courtesy of BioServe.*

Students Participating in ISS Investigator Experiments. NASA Image ISS015E10587 View of Capillary Flow Experiment (CFE) in the U.S. Laboratory/Destiny. Photo was taken during Expedition 15. *Image courtesy of NASA.*

Students Participating in ISS Engineering Education: Hardware Development. Three satellites fly in formation as part of the Synchronized Position Hold, Engage, Reorient, Experimental Satellites (SPHERES) investigation. *Image courtesy of NASA.*

Educational Demonstrations and Activities. A student talks to a crewmember onboard the ISS during an ARISS contact. *Image courtesy of ARISS.*

Cultural Activities. Elementary school attached to the Faculty of Education, Shimane University (Matsue City, Shimane Prefecture, Japan) and the Japan Aerospace Exploration Agency (JAXA). *Image courtesy of JAXA.*

International Space Station Education Accomplishments

Student-Developed Investigations. West Ward Elementary students participate in an experiment as Astronaut Daniel Burbank and West Ward Principal Maureen Adams observe.
Education Competitions. NASA Explorer Schools 2009 Student Symposium.

Students Performing Classroom Versions of ISS Investigations. Female *Nephila clavipes* (golden orb spider) in her web. *Image taken by Danielle Anthony.*

Students Participating in ISS Investigator Experiments. NASA Image ISS006E26867. View of a surface tension demonstration using water that is being held in place by a 50mm metal loop. *Image courtesy of NASA.*

Students Participating in ISS Engineering Education: Hardware Development. Close up of the experiment trays in MISSE, open and exposed to space. *Image courtesy of NASA.*

Educational Demonstrations and Activities. View of surface tension demonstration during Saturday Morning Science, using water that is being held in place by a metal loop. Food coloring has been added to the water for demonstration purposes only. *Image courtesy of NASA.*

Cultural Activities. Image of the Moon from ISS, used as inspiration for the musical score for the Moon Score activity. *Image courtesy of NASA.*

Contents

Introduction *1*

    Summary of International Space Station Education Opportunities and Accomplishments 5

    Summary of Inquiry-Based International Space Station Education 6

International Space Station Education Opportunities *7*

    Student-Developed Investigations..*9*

        Earth Knowledge Acquired by Middle School Students (EarthKAM) 11

        Drop Your Thesis! 16

        Fly Your Thesis! 17

        NanoRacks NanoLabs..19

        NanoRacks-National Center for Earth and Space Science Education-1, -2, and -Aquarius (NanoRacks-NCESSE-1, -2, and -Aquarius)..21

        Reduced-Gravity Education Flight Program 25

        Rocket and Balloon Experiments for University Students (REXUS/BEXUS) 27

        JAXA Seeds in Space I (Asagao and Miyako-Gusa)..28

        JAXA Spaceflight Seeds Kids I (Himawari) 30

        Take Your Classroom Into Space (TYCIS) 32

        Tomatosphere-III 35

    Education Competitions..*37*

        Kids in Micro-g...39

        Spaced Out Sports...43

        Synchronized Position Hold, Engage, Reorient, Experimental Satellites (SPHERES) – Zero Robotics..45

        YouTube Space Lab..48

    Students Performing Classroom Versions of ISS Investigations *51*

        Commercial Generic Bioprocessing Apparatus Science Insert – 05: Spiders, Fruit Flies and Directional Plant Growth (CSI-05) 53

    Students Participating in ISS Investigator Experiments..*55*

        Crew Earth Observations (CEO) 57

        Student Involvement in Other ISS Investigations..60

        Analysis of a Novel Sensory Mechanism in Root Phototropism (Tropi) 62

        Binary Colloidal Alloy Test-3 and -4: Critical Point (BCAT-3-4-CP)................................63

Capillary Channel Flow (CCF) 63

Capillary Flow Experiments (CFE) 64

Cardiovascular and Cerebrovascular Control on Return From International Space
Station (CCISS); Cardiovascular Health Consequences of Long-Duration
Space Flight (Vascular) 65

Constrained Vapor Bubble (CVB) 65

Coulomb Crystal 66

Device for the Study of Critical Liquids and Crystallization – Directional
Solidification Insert (DECLIC-DSI) 67

ELaboratore Immagini TElevisive – Space 2 (ELITE-S2) 68

Hyperspectral Imager for Coastal Ocean (HICO) and Remote Atmospheric and
Ionospheric Detection System (RAIDS) Experiment Payload – Remote Atmospheric
and Ionospheric Detection System (HREP-RAIDS) 68

Integrated Resistance and Aerobic Training Study (SPRINT) 69

Investigating the Structure of Paramagnetic Aggregates from Colloidal Emulsions–2
(InSPACE-2) 70

Materials International Space Station Experiment – 7 (MISSE-7) 70

Mental Representation of Spatial Cues During Space Flight (3D-Space) 70

Microheater Array Boiling Experiment (MABE): Flight Research Using the Boiling
Experiment Facility (BXF) 71

Nucleate Pool Boiling Experiment (NPBX) 71

Nutritional Status Assessment (Nutrition) and Dietary Intake Can Predict and Protect
Against Changes in Bone Metabolism during Spaceflight and Recovery (Pro K) 72

Physiological Factors Contributing to Changes in Postflight Functional Performance
(Functional Task Test) 73

Psychomotor Vigilance Self Test on the International Space Station
(Reaction Self Test) 73

Sensor Test for Orion Relative Navigation Risk Mitigation (STORRM) 74

Space Communications and Navigation Testbed (SCAN Testbed) 74

Space Dynamically Responding Ultrasonic Matrix System (SpaceDRUMS) 75

Students Participating in ISS Engineering Education: Hardware Development 77

High School Students United With NASA to Create Hardware (HUNCH) 79

International Space Station Agricultural Camera (ISAAC) 81

Polymers Erosion and Contamination Experiment (PEACE) and Polymers Experiments
on MISSE 2, 5, 6, 7, and 8 83

Robonaut 85

Synchronized Position Hold, Engage, Reorient, Experimental Satellites (SPHERES)..........86

Educational Demonstrations and Activities..89

Butterflies, Spiders and Plants in Space 91

Education Payload Operations – Demonstrations (EPO-DEMOS)..........................93

JAXA Education Payload Operations – Demonstrations (JAXA EPO–Demos)......99

ESA – Mission Digital Video Disk (DVD) Series..103

ESA – Space in Bytes Series 105

Global Water Experiment 106

International Space Station In-flight Education Downlinks (In-flight Education Downlinks)..108

International Space Station Ham Radio (ISS Ham Radio) [Also Known as Amateur Radio on the International Space Station (ARISS)] 110

International Space Station *Live!* (ISS*Live!*) 115

International Toys in Space 117

Lego® Bricks..118

Mission X: Train Like an Astronaut 120

Space Devices and Modern Technology for Personal Communication (MAI-75).122

Ten-Mayak (Shadow-Beacon)..123

Try Zero-Gravity (Try Zero-G) 125

Cultural Activities..*127*

Great Start 129

JAXA Uchu Renshi (Space Poem Chain)...130

Summary of International Space Station Education Opportunities 132

International Space Station Education Accomplishments..***133***

Student-Developed Investigations..*135*

Analysis of Inertial Solid Properties (APIS)...137

Biological Research in Canisters – 16: Actin Regulation of Arabidopsis Root Growth and Orientation During Space Flight (BRIC-16-Regulation)................................137

Drop Your Thesis! 2011: Falling Roots...138

Drop Your Thesis! 2010: Bubble Jet Impingement in Microgravity Conditions 139

Drop Your Thesis! 2009: New Polymer Dispersed Liquid Crystal (PDLC) Materials Obtained From Dispersion of Liquid Crystal in Microgravity Conditions............140

Educational Demonstration of Basic Physics Laws of Motion (Fizika-Obrazovanie).........141

Electrostatic Self-Assembly Demonstration (ESD) 143

Fly Your Thesis! 2009: ABC Transporters in Microgravity 144

Fly Your Thesis! 2009: AstEx – Simulating Asteroidal Regoliths: Implications for Geology and Sample Return 145

Fly Your Thesis! 2009: Complex — Microgravity Studies of the Effect of Volume Fraction and Salinity on Flow in Samples of Clay Nanoparticles Dispersed in Water 146

Fly Your Thesis! 2009: Dust Side of the Force — The Importance of the Temperature Gradient Effect in Planet Formation Processes and Dust Devils or Storms on Mars 147

Foam Optics and Mechanics – Stability (Foam-Stability) 148

Image Reversal in Space (Iris) 149

Space Experiment Module (SEM) 149

Test of the Basic Principles of Mechanics in Space (THEBAS) 150

University Research Centers–Microbial-1 (URC-Microbial-1) 151

Education Competitions *139*

SUCCESS Programme 155

SUCCESS Bug Energy — Study of Output of Bacterial Fuel Cells in Weightlessness 155

SUCCESS Chondro – Study on the Development of Methods To Produce Artificial Cartilage 156

SUCCESS Graphobox – Study Into the Interaction of Effects of Light and Gravity on the Growth Processes of Plants 156

SUCCESS UTBI – Under the Background Influence 157

SUCCESS Winograd – Effects of Gravity on Bacterial Development 157

Students Performing Classroom Versions of ISS Investigations *139*

Advanced Astroculture (ADVASC) and Photosynthesis Experiment and System Testing and Operation (PESTO) 161

Agrospace Experiment Suite (AES) 161

Brazilian Seeds *Phaseolus Vulgaris*: Demonstration of Gravitopism and Phototropism Effects on the Germination of Seed in Microgravity (SED) 162

Commercial Generic Bioprocessing Apparatus Science Insert–01: *C. Elgans* and Seed Germination (CSI-01) 163

Commercial Generic Bioprocessing Apparatus Science Insert–02: Silicate Garden, Seed Germination, Plant Cell Culture and Yeast (CSI-02) 164

Commercial Generic Bioprocessing Apparatus Science Insert–03: Spiders and Butterflies (CSI-03) 165

Education Payload Operations – Kit C: Plant Growth Chambers (EPO-Kit C) 167

Educational Demonstration of the Effects of Shape and Size on the Recovery of Precompressed Plastic Material (MATI-75) 168

Farming In Space – Biomass Production System (BPS) 169

Oil Emulsion Experiment (OEE) 170

Seeds in Space (Seeds) 171

Silkworms in Space – Kinugusa-Kai (Radsilk) 171

Space-Exposed Experiment Developed for Students (Education-SEEDS) 172

Students Participating in ISS Investigator Experiments 161

Advanced Diagnostic Ultrasound in Microgravity (ADUM) 175

Coarsening in Solid Liquid Mixtures-2 (CSLM-2) 176

Commercial Biomedical Test Module – 2 (CBTM-2) 176

Commercial Generic Bioprocessing Apparatus – Antibiotic Production in Space (CGBA-APS) 177

Crewmember and Crew-Ground Interaction During International Space Station Missions (Interactions) 177

Differentiation of Bone Marrow Macrophages in Space (BONEMAC) 178

Fluid Merging Viscosity Measurement (FMVM) 178

International Space Station Zero-Propellant Maneuver (ZPM) Demonstration 179

Intravenous Fluid Generation Experiment (IVGEN) 180

Italian Foam (I-Foam) 181

Mice Drawer System (MDS) 182

Mouse Antigen-Specific CD4+ T Cell Priming and Memory Response During Spaceflight (Mouse Immunology) 183

National Laboratory Pathfinder – Cells – *Jatropha* Biofuels (NLP-Cells-3, 4, 6, 7) 183

Passive Observatories for Experimental Microbial Systems (POEMS) 184

Pore Formation and Mobility During Controlled Directional Solidification in a Microgravity Environment (PFMI) 185

Sleep-Wake Actigraphy and Light Exposure During Spaceflight-Long (Sleep-Long) and Sleep-Wake Actigraphy and Light Exposure During Spaceflight-Short (Sleep-Short) .. 185

Space Flight Induced Reactivation of Latent Epstein-Barr Virus (Epstein-Barr) 186

Student Access to Space (Part of Protein Crystal Growth-Enhanced Gaseous Nitrogen Dewar) (PCG-EGN) 187

Students Participating in ISS Engineering Education: Hardware Development 173

Filming of Space Robot "Jitter" Assembled Out of LEGOS (Konstructor) 191

    Kolibry 191

Educational Demonstrations and Activities *173*

    Demonstration of Mass and Weight of Objects and Action of Reactive Forces in Microgravity (Education-SA) 195

    Electronic-Learning (E-Learning) 195

    Education — How Solar Cells Work (Education-Solar Cells) 196

    Greenhouse in Space (ESA-EPO-Greenhouse) 196

    Kibo Kids Tour 198

    Science of Opportunity (Saturday Morning Science) 198

    Taste in Space 200

Cultural Activities *173*

    DREAMTiME (DREAMTiME) 203

    Get Fit for Space Challenge with Bob Thirsk (Get Fit for Space) 203

    Suit Satellite (RadioSkaf) 204

Summary of International Space Station Education Accomplishments 206

Appendix: International Space Station Missions *207*

Tables

Table 1 – Student, Teacher and School Summary of ISS Education Opportunities and Accomplishments *5*

Table 2 – Inquiry-Based Student, Teacher and School Summary of ISS Education Opportunities and Accomplishments *6*

Table 3 – Countries and Number of Schools That Participated in EarthKAM *14*

Table 4 – Number of Schools by State That Participated in EarthKAM *15*

Table 5 – NanoRacks-NCESSE-Aquarius Investigations *22*

Table 6 – NanoRacks-NCESSE-2 Investigations *23*

Table 7 – NanoRacks-NCESSE-1 Investigations *24*

Table 8 – ESA Education Payload Operations and Demonstrations *34*

Table 9 – Kids in Micro-g List of Winners *41*

Table 10 – Kids in Micro-g-2 List of Winners *42*

Table 11 – Student Involvement in Other International Space Station Investigations *60*

Table 12 – NASA Education Payload Operations – Demonstrations *94*

Table 13 – JAXA Education Payload Operations – Demonstrations *100*

Table 14 – ESA DVD and Educational Kits *104*

Table 15 – Countries and Number of Schools That Conducted ARISS Contacts *113*

Table 16 – ARISS Contacts by U.S. States *114*

Table 17 – Countries and Number of Students Participating in Mission X: Train Like an Astronaut *121*

Table 18 – Student, Teacher and School Summary of ISS Education Opportunities *132*

Table 19 – Number of Teachers by State That Participated in Butterflies in Space *166*

Table 20 – Student, Teacher and School Summary of ISS Education Accomplishments *206*

Table 21 – International Space Station Expeditions and Crewmembers *209*

Table 22 – Shuttle Flights to the International Space Station *212*

Table 23 – Soyuz Flights to the International Space Station *215*

Table 24 – International Space Station Progress Supply Missions *216*

Table 25 – Automated Transfer Vehicle (ATV) *217*

Table 26 – H-II Transfer Vehicle (HTV) *217*

Acronyms and Abbreviations

ABC	ATP Binding Cassette
ADUM	Advanced Diagnostic Ultrasound in Microgravity
ADVASC	Advanced Astroculture
AgCam	Agricultural Camera
APIS	Analysis of Inertial Solid Properties
ARISS	Amateur Radio on International Space Station
ASI	Italian Space Agency
ATV	Automated Transfer Vehicle
BCAT-3-CP	Binary Colloidal Alloy Test-3-Critical Point
BCM	Baylor College of Medicine
BDC	Base Data Collection
BXF	Boiling Experiment Facility
CBTM	Commercial Biomedical Test Module
CCF	Capillary Channel Flow
CCISS	Cardiovascular and Cerebrovascular Control on return from International Space Station
CEO	Crew Earth Observations
CFE	Capillary Flow Experiments
CGBA-APS	Commercial Generic Bioprocessing Apparatus – Antibiotic Production in Space
CORE	Central Operation of Resources for Educators
CSA	Canadian Space Agency
CSI	Commercial Generic Bioprocessing Apparatus Science Insert
CSLM	Coarsening in Solid Liquid Mixtures
CVB	Constrained Vapor Bubble
DARPA	Defense Advanced Research Projects Agency
DECLIC-DSI	Device for the Study of Critical Liquids and Crystallization – Directional Solidification Insert
DI	Deionizing
DLN	Digital Learning Network

DLR	German Aerospace Center
DRUMS	(Space)-Dynamically Responding Ultrasound Matrix System
EarthKAM	Earth Knowledge Acquired by Middle School Students
ELGRA	European Low Gravity Research Association
ELITE-S2	ELaboratore Immagini TElevisive – Space 2
EPO	Education Payload Operations
EPO-Demos	Education Payload Operations Demonstrations
ESA	European Space Agency
ESD	Electrostatic Self-assembly Demonstration
ESTEC	European Space Research and Technology
EVA	Extravehicular Activity
FCA	Faith Christian Academy
FCHS	Fremont Christian High School
FCUP	Faculdade de Ciências da Universidade do Porto
FIRST	For Inspiration and Recognition of Science and Technology
FME	Fluid Mixing Enclosure
FMVM	Fluid Merging Viscosity Measurement
GAPE	Gateway to Astronaut Photography of Earth
GRA	Graduate Research Assistantship
GRC	Glenn Research Center
HREP	Hyperspectral Imager for Coastal Ocean (HICO) and Remote Atmospheric and Ionospheric Detection System (RAIDS) Experiment Payload
HTV	H-II Transfer Vehicle
HUNCH	High School Students United With NASA To Create Hardware
InSPACE	Investigating the Structure of Paramagnetic Aggregates from Colloidal Emulsions
ISAAC	International Space Station Agricultural Camera
ISCED	International Standard Classification for Education
ISS	International Space Station
IUPAC	International Union of Pure and Applied Chemistry
JAXA	Japan Aerospace Exploration Agency

JEM	Japanese Experiment Module
JSC	Johnson Space Center
KSC	Kennedy Space Center
LED	Light-Emitting Diodes
MABE	Microheater Array Boiling Experiment
MDA	Materials Diffusion Apparatus
MDS	Mice Drawer System
MISSE	Materials International Space Station Experiment
MIT	Massachusetts Institute of Technology
MORABA	Mobile Rocket Base
MOVE	Movement in Orbital Vehicle Experiments
MR	Magnetorheological
MRP2	Multidrug Resistance-associated Protein 2
NASA	National Aeronautics and Space Administration
NCESSE	National Center for Earth and Space Science Education
NLP	National Laboratory Pathfinder
NPBX	Nucleate Pool Boiling Experiment
NSO	Netherlands Space Office
OSF	Osteoblast Stimulating Factor
PC	Personal Computer
PCG-EGN	Protein Crystal Growth-Enhanced Gaseous Nitrogen
PDLC	Polymer Dispersed Liquid Crystals
PEACE	Polymers Erosion and Contamination Experiment
POEMS	Passive Observatories For Experimetnal Microbial Systems
PVT	Payload Verification Testing
RAIDS	Remote Atmospheric and Ionospheric Detection System
REU	Research Experiences for Undergraduates
Roscosmos	Russian Federal Space Agency
RS	Russian Segment
SCAN	Space Communication and Navigation (Testbed)
SDR	Software Defined Radio

SEEDS	Space Exposed Experiment Developed for Students
SEM	Space Experiment Module
SNSB	Swedish National Space Board
SPHERES	Synchronized Position, Hold, Engage, Reorient Experimental Satellites
SPRINT	Integrated Resistance and Aerobic Training Study
SSC	Swedish Space Corporation
SSEP	Student Spaceflight Experiment Program
SSL	Space Systems Laboratory
STEM	Science, Technology, Engineering, and Mathematics
STORRM	Sensor Test for Orion Relative Navigation Risk Mitigation
SVT	Science Verification Testing
TROPI-2	Functional Gravity Studies in Sensory Mechanism in Root Phototropism
TYCIS	Take Your Classroom Into Space
UCSD	University of California at San Diego
UND	University of North Dakota
UNESCO	United Nations Educational, Scientific and Cultural Organization
UPC	Universitat Politècnica de Catalunya
URC	University Research Centers
UTMB	University of Texas Medical Branch
UV	Ultraviolet
VCHS	Valley Christian High School
VHF	Very High Frequency
WCHS	Whittier Christian High School
WORF	Window Observational Research Facility
WPA	Water Processor Assembly
ZPM	Zero-Propellant Maneuver

Foreword

The International Space Station (ISS) is without a doubt history's most sophisticated engineering marvel. It has become synonymous with many terms — space exploration, ingenuity, collaboration, inspiration, and even sacrifice. The ISS' accomplishments are countless in regard to scientific and space exploration research. However, one aspect that is sometimes overlooked is its contributions to education, particularly STEM (science, technology, engineering, and mathematics) education.

As a NASA astronaut and engineer, I have followed the progress of the ISS. It has been fascinating to see how the place I called home for 23 days has been inspiring students of all ages to pursue science and engineering degrees. Many of these students have a goal of becoming one of the next ISS crew members or the first person to walk on Mars. The greatest impression that living on the space station left on me was the remarkable brilliance and strength of the human mind, and how this masterpiece would leave an everlasting footprint in global human endeavors. It is that same footprint today which ignites children's imagination and curiosity for science and space, and will continue to do so for generations to come.

Space station educational activities have had an inspirational impact on millions of students worldwide, by involving them in station research, and by using the station to teach them the science and engineering that are behind space exploration. In the first twelve years of continuous human presence on the space station, a wide range of experiments for students of all ages have been performed thanks to astronauts and cosmonauts from 15 different nations. I performed experimental activities with *Escherichia coli* (*E. coli*) bacteria for graduate students and also returned mice to our planet, after their stay on orbit looking at the micro-gravity effects. Other experiments included observing how tomato seeds that have been flown to space grow differently on Earth; speaking live with ISS astronauts through station downlinks; and having students design satellites that are tested aboard the ISS, among several others.

The ISS has vividly increased the public standing of science for students, educators, and world citizens. At a time when technology and science are leading societies, we must highlight their roles in STEM education, so that we can continue inspiring tomorrow's scientists, engineers, and mathematicians. The educational opportunities brought to students worldwide, have been possible thanks to the collaborations between the National Aeronautics and Space Administration, the Russian Federal Space Agency, the Japan Aerospace Exploration Agency, the European Space Agency, and the Canadian Space Agency. It is no wonder several of the experiments developed for students have also provided them with cultural experiences, linking science with the humanities.

The experiments, activities, and photographs on the following pages exemplify the opportunities and accomplishments delivered by the ISS. It is our planet's most unique educational resource for students and educators around the globe. This is merely a glimpse of how the ISS has sparked the imagination and interest of thousands of students, who dream to be the next generation's space explorers.

Leland Melvin
Associate Administrator for Education and NASA Astronaut

Executive Summary

This publication summarizes the main student experiments and educational activities that have been performed on board the International Space Station (ISS) from its first element launch in the year 2000 to 2011. It begins with active opportunities that are available to students today and includes a summary of activities and participants from the past. This is a comprehensive account of the education projects that have been conducted and led by the ISS Partners — National Aeronautics and Space Administration (NASA), Canadian Space Agency (CSA), European Space Agency (ESA), Japan Aerospace Exploration Agency (JAXA) and Russian Federal Space Agency (Roscosmos).

While the immediate impact that these programs have had on our students is understandably challenging to measure, there are a number of examples of students who have pursued advanced education in technical areas as a result of their participation in these programs and their exposure to the excitement and wonder of space exploration.

The projects summarized in this publication represent the enthusiasm to use the ISS as an educational platform. This enthusiasm is shared by university researchers, commercial companies, non-governmental organizations, other international government agencies, students and educators themselves. If the last 12 years of diverse and successful educational activities are any indication of the interest in and excitement for using this unique resource, the era of utilization promises to be even more successful. It is clear, based on the student experiments and activities already completed, the ISS has the exceptional ability to inspire the next generation of thinkers, innovators, leaders and explorers in science, engineering and technology. They will be well prepared to meet the challenges of tomorrow.

Third graders from Sacaton, Arizona, tracking the International Space Station for their EarthKAM project. Image courtesy of Sally Ride Science EarthKAM.

Introduction

The International Space Station (ISS) has a unique ability to capture the imaginations of both students and teachers worldwide. For the past 12 years, the presence of humans on board the ISS has provided a foundation for numerous educational activities aimed at capturing that interest and motivating study in science, technology, engineering and mathematics (STEM).

Projects such as the Amateur Radio on the International Space Station (ARISS) and Earth Knowledge Acquired by Middle School Students (EarthKAM) as well as events like Take Your Classroom Into Space (TYCIS) have allowed for global student, teacher and public access to space through student image acquisition and radio contacts with crewmembers. Educational activities are not limited to STEM, as illustrated by the Uchu Renshi project, in which a chain poem initiated by a crewmember in space is continued and completed by people on Earth. With ISS operations continuing at least until 2020, such projects and the educational materials that accompany them will be made available to more students around the world.

It was recognized very early in the program that students would have a strong interest in the ISS and that they would be provided a unique opportunity to participate in science and engineering projects on the ISS. Since the launch of first element of the ISS, student experiments and educational activities have been performed by all of the international partner countries officially participating in the ISS Program and several countries that participate under commercial agreements. Many of these early programs still continue, and others are being developed and added regularly. Each of these diverse student experiments and programs falls into one of the following categories.

Student-Developed Investigations. These experiments, which are performed by students under the aegis of a teacher or scientist mentor, are created solely for the benefit of the students. One example of this is the Synchronized Position, Hold, Engage, Reorient Experimental Satellites (SPHERES) – Zero-Robotics competition in which high school students write algorithms for the SPHERES satellites to accomplish tasks relevant to future space missions.

The International Space Station photographed by a crewmember aboard the space shuttle.
Image S119E010316

A fifth-grade student from Skinner West Classical, Fine Arts and Technology School in Chicago, Illinois, whose experiment to determine the effects of microgravity on the development of goldfish was selected for flight as a part of the Student Spaceflight Experiments Program (SSEP). Image courtesy of the National Center for Earth and Space Science

Education Competitions. These educational activities involve a student design competition, and students usually have an opportunity to submit experiment proposals that will be assessed and judged. Crewmembers often conduct the experiments proposed by the winners of the competition on the ISS.

Used as an educational platform, the ISS affords students limitless motivational and inspirational opportunities. The programs have allowed participation with students at all grade levels — from kindergarten through twelfth grade as well as undergraduate and graduate students and postdoctoral fellows.

Students Performing Classroom Versions of ISS Investigations. These experiments, performed by students in their classrooms, mimic experiments that are being conducted or have been conducted by professional researchers on the ISS. These experiments provide an opportunity for students to observe differences between their results and results from experiments being performed by crewmembers on the ISS. A typical example of this type of experiment is Tomatosphere-III, where students measure the germination rates, growth patterns and vigor of growth of seeds flown on the ISS and compare these to the same values for seeds grown on Earth.

Students Participating in ISS Investigator Experiments. Many ISS investigators have enlisted students to help them with their experiments. Some experiments are performed solely to inspire the next generation of explorers. Others involve undergraduate and graduate students as well as postdoctoral fellows who are working under the guidance of investigators and university professors. An example of this category is Capillary Channel Flow (CCF), which is a research investigation, but involves students in the performance, evaluation and computation of the experiment. High school students participating in the Materials International Space Station Experiment (MISSE)

During a previous Tomatosphere program, students studied the growth of tomato plants in Miss Smith's grade three class at Langley Fundamental Elementary, Vancouver, British Columbia, Canada. The students took their plants home to grow in their gardens over the summer. Image courtesy of Tomatosphere.

Polymers Erosion and Contamination Experiment (PEACE) have received national science fair awards for their work.

Students Participating in ISS Engineering Education: Hardware Development. These activities typically involve high school and college students developing hardware to support the ISS Program as they learn about and participate in space operations. The Agricultural Camera (AgCam) — a project that ran from 2005 to 2010 — is one example in which undergraduate and graduate students from the University of North Dakota designed, built and operated ISS experiment hardware.

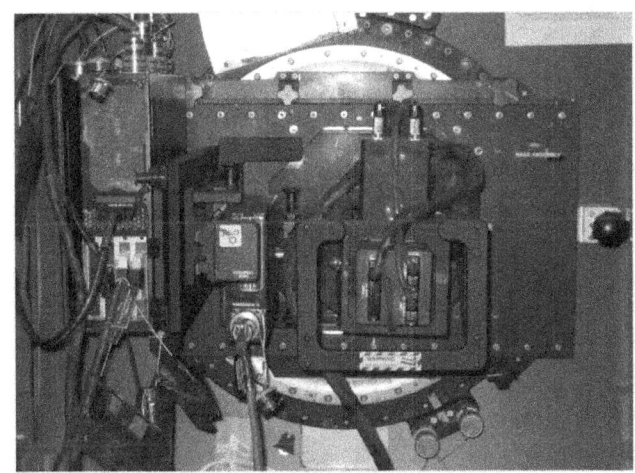

AgCam hardware integrated on the window in the Destiny Laboratory mockup at the Johnson Space Center in Houston, Texas. Image courtesy of NASA.

Educational Demonstrations and Activities. In addition to the numerous ISS experiments supported by students, several other educational activities and demonstrations are performed to be used as teaching aids, to supply resource materials, or simply to provide additional mechanisms to inspire students. These range from crewmembers demonstrating how simple and familiar phenomena, such as static attraction in microgravity, reveal an exciting new understanding of physics in space to allowing students of all ages to pose questions to the crew on board the station. These diverse activities are rich with opportunities to connect with students and bring the ISS experience into their lives.

Students plant seeds of Himawari. (Kotobuki-kita Elementary School, Kanoya City, Kagoshima Prefecture, Japan). Image courtesy of Kotobuki-kita Elementary School and JAXA.

Cultural Activities. Along with educational activities that have a technical component, there are activities that provide a cultural experience to the students. These include various cultural activities, such as dancing in microgravity or playing an instrument, performed by crewmembers on board the ISS. On the ground, students engage in such cultural experiences as the Japanese Himawari, also known as JAXA Spaceflight Seeds Kids I. Students in Japan had the opportunity to grow sunflower seeds that flew on board the station for approximately nine months in 2008 as part of Touch the Universe. Through a series of education guides, students gained a deeper understanding of the past and near-term future of human space flight as well as the future of Earth and life.

Student scientists have an opportunity to design and fly their own experiments on the ISS, such as the fifth graders (10 and 11 year olds) who, as part of the National Center for Earth and

Space Science Education (NCESSE) Student Spaceflight Experiments Program (SSEP) working with the U.S. commercial company NanoRacks as their implementation partner, investigated the effect of microgravity on the development of goldfish eggs. Countless high school students involved in ISS projects have steered their careers to mathematics, science and other technical areas. Many college students also have been involved in ISS research and have completed various projects, theses and dissertations related to ISS experiments.

In addition to inspiring students, ISS education programs also motivate teachers. Some of the station experiments have encouraged teachers to participate through workshops that facilitate them in leading students to design and conduct ISS experiments. By inspiring the teachers, the impact of the ISS experience can become even broader. Teachers have demonstrated that they serve as a great resource for promoting enthusiasm for space exploration through implementing their training in the classroom.

ISS student experiments and activities have involved schools from all over the world. To date, more than 43 million students from 49 countries have participated. Students from all across Asia, Australia, Europe, Africa, and North and South America, including the Caribbean region, have had opportunities to be inspired by their involvement in ISS education activities and as a result have been motivated to pursue careers in STEM. Projects such as EarthKAM and ARISS have made it possible for students in some of the most remote places in the world to communicate with and become involved in the ISS Program.

Summary of International Space Station Education Opportunities and Accomplishments

Table 1 – Student, Teacher and School Summary of ISS Education Opportunities and Accomplishments

Investigations	Number of Students				Schools	Teachers
	K-12	Undergraduate	Graduate	Postdoctoral		
Student-Developed Investigations	560,612	347	9		17,293	19,308
Education Competitions	3806				74	77
Students Performing Classroom Versions of ISS Investigations	1,059,938	37	28		5078	5287
Students Participating in ISS Investigators Experiments	89,768	1422	1167	47	66	6255
Students Participating in ISS Engineering Education: Hardware Development	2116	273	46		6	63
Educational Demonstrations and Activities	41,480,680				3437	2,797,620
Cultural Activities	620				15	20
Totals	~43,000,000	~2,000	~1,200	~50	~26,000	~2,800,000

Summary of Inquiry-Based International Space Station Education

Inquiry-based education is an investigative approach to learning that encourages students, individually and collaboratively, to observe, ask questions, design inquiries, and collect and interpret data in order to develop concepts and relationships from empirical experiences. (Standards for Science Teacher Preparation, NSTA 2003).

Table 2 – Inquiry-Based Student, Teacher and School Summary of ISS Education Opportunities and Accomplishments

Investigations	Number of Students				Schools	Teachers
	K-12	Undergraduate	Graduate	Postdoctoral		
Student-Developed Investigations	560,612	347	9		17,293	19,308
Education Competitions	3806				74	77
Students Performing Classroom Versions of ISS Investigations	1,059,938	37	28		5078	5287
Students Participating in ISS Investigators Experiments	89,768	1422	1167	47	66	6255
Students Participating in ISS Engineering Education: Hardware Development	2116	273	46		6	63
Totals	**~1,700,000**	**~2,000**	**~1,200**	**~50**	**~23,000**	**~31,000**

International Space Station Education Opportunities

These educational activities and projects are ongoing and serve as opportunities for students at various levels to be involved.

Student-Developed Investigations

These experiments, which are typically developed by students under the aegis of a teacher or scientist mentor, are performed solely for the benefit of the students.

Earth Knowledge Acquired by Middle School Students (EarthKAM)

Expeditions: 2–32, Ongoing

Leading Space Agencies: NASA, Global

Curriculum Grade Levels: K–8 (elementary)

Impact: Approximately 2800 schools with 190,000 middle school students and 3000 teachers in the United States and 48 other countries have participated in EarthKAM. A total of 150 undergraduate students from the University of California at San Diego (UCSD), San Diego, California, also have participated in integrating and operating the experiment.

Participating Countries: Argentina, Australia, Belgium, Bermuda, Bolivia, Brazil, Canada, Chile, China, Columbia, Croatia, Czech Republic, Denmark, Dominican Republic, Egypt, England, Fiji, Finland, France, Germany, Ghana, Greece, Guatemala, India, Italy, Japan, Kenya, Lebanon, Malaysia, Mexico, Netherlands, New Zealand, Nigeria, Poland, Polynesia, Portugal, Puerto Rico, Romania, Russia, South Africa, South Korea, Spain, Switzerland, Trinidad/Tobago, Turkey, United Kingdom, United States, Uruguay, Venezuela

Participating States: Alabama, Alaska, Arizona, Arkansas, California, Colorado, Connecticut, Delaware, Florida, Georgia, Hawaii, Idaho, Illinois, Indiana, Iowa, Kansas, Kentucky, Louisiana, Maine, Maryland, Massachusetts, Michigan, Minnesota, Mississippi, Missouri, Montana, Nebraska, Nevada, New Hampshire, New Jersey, New Mexico, New York, North Carolina, North Dakota, Ohio, Oklahoma, Oregon, Pennsylvania, Rhode Island, South Carolina, South Dakota, Tennessee, Texas, Utah, Vermont, Virginia, Washington, West Virginia, Wisconsin, Wyoming

Number of K–8 Students (elementary): 190,154

Number of Undergraduate Students (college, postsecondary): 150

Number of Teachers: 3000

Number of Schools: 2872

Ongoing Opportunities: Approximately 60 Schools per Year

> *Marvin Singer of Francis Scott Key School, Philadelphia, Pennsylvania, USA: "I've told my students that the EarthKAM experience is one they will long remember and should be included in their academic resume."*
>
> *Tetsuya Fukuda of Nara Education University JHS in Japan: "My students tried the EarthKAM project this April. It's wonderful. My students could get a great experience about space education."*
>
> *Maria Alexis of Pope John Paul II Regional Catholic Elementary School in West Brandywine Township, Pennsylvania: "Our fourth graders truly enjoyed working together on this mission and learning about the geography of our Earth and the roles satellites and the ISS play in providing information to scientists around the world."*

Description of Student Participation and Activities: EarthKAM is a NASA-sponsored education program that enables thousands of students to photograph and examine the Earth from the unique perspective of space. Using the EarthKAM website, students control a special, digital camera mounted in a window on the ISS. From this camera, students are able to photograph a wide range of beautiful and fascinating features on the surface of Earth. During an EarthKAM session, middle school students select photographic targets linked to the curriculum of each school. Undergraduate students at UCSD integrate the requests from the schools at the Mission Operations Center and send a camera control file to the Johnson Space Center (JSC) in Houston, Texas, for uplink to the ISS. The resulting photographs are downlinked from the station back to EarthKAM and are made available on the Web to be viewed and shared by participating classrooms and the general public.

Description of Teacher Participation and Activities: The role of the teachers is to help facilitate the experience described above for their students. They also help develop related curricula and lesson plans related to the studies of their students.

Principal Investigator: Sally Ride, Ph.D., Sally Ride Science

Education Lead: Teresa Sindelar, Teaching From Space, NASA JSC, Houston, Texas, USA

Education Websites:
http://www.nasa.gov/education/tfs
http://earthkam.ucsd.edu

Students at Gonzalez Middle School in Donaldsonville, Louisiana, are using computers to help select potential target sites for the EarthKAM camera. Image courtesy of Sally Ride Science EarthKAM.

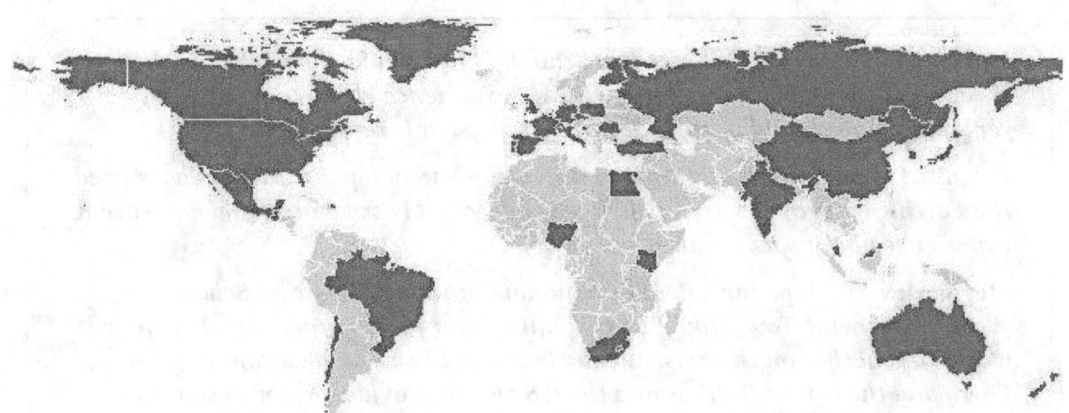

Map of countries that participated in EarthKAM missions.

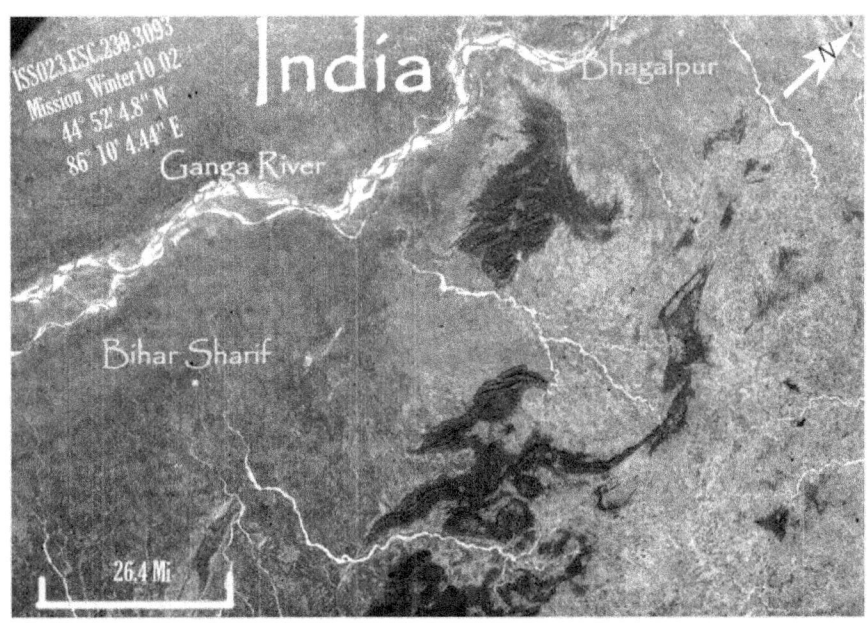

Taken during the EarthKAM Winter 2010 mission, this image of Bihar, India, shows the variation in terrain as well as the Ganga River. The image was annotated by an EarthKAM intern at UCSD.

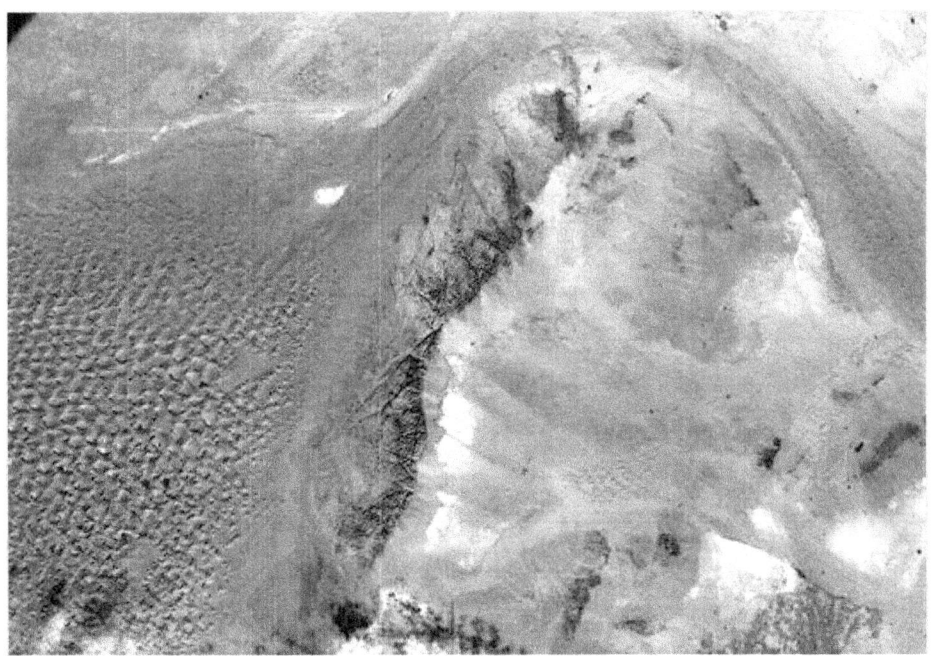

This image was taken during the EarthKAM Winter 2010 mission. The Yabrai Shan Mountains of China are depicted along with surrounding hills and desert. The image was annotated by an EarthKAM intern at UCSD.

Table 3 – Countries and Number of Schools That Participated in EarthKAM

Country	No. of Schools	Country	No. of Schools
Argentina	22	Japan	59
Australia	35	Kenya	1
Belgium	1	Lebanon	5
Bermuda	1	Malaysia	3
Bolivia	1	Mexico	7
Brazil	2	Netherlands	5
Canada	33	New Zealand	3
Chile	6	Nigeria	2
China	2	Poland	22
Columbia	6	Polynesia	2
Croatia	5	Portugal	4
Czech Republic	2	Puerto Rico	8
Denmark	2	Romania	11
Dominican Republic	7	Russia	1
Egypt	3	South Africa	1
England	4	South Korea	7
Fiji	2	Spain	25
Finland	2	Switzerland	3
France	14	Trinidad and Tobago	1
Germany	30	Turkey	5
Ghana	1	United Kingdom	11
Greece	1	United States	2222
Guatemala	2	Uruguay	1
India	53	Venezuela	1
Italy	10		

Table 4 – Number of Schools by State That Participated in EarthKAM

State	No. of Schools	State	No. of Schools
Alabama	20	Montana	7
Alaska	16	Nebraska	14
Arizona	39	Nevada	18
Arkansas	24	New Hampshire	12
California	235	New Jersey	70
Colorado	23	New Mexico	20
Connecticut	35	New York	104
Delaware	7	North Carolina	60
Florida	207	North Dakota	2
Georgia	57	Ohio	68
Hawaii	4	Oklahoma	31
Idaho	14	Oregon	7
Illinois	50	Pennsylvania	102
Indiana	31	Rhode Island	10
Iowa	16	South Carolina	29
Kansas	29	South Dakota	5
Kentucky	19	Tennessee	6
Louisiana	27	Texas	398
Maine	7	Utah	21
Maryland	62	Vermont	9
Massachusetts	50	Virginia	55
Michigan	41	Washington	35
Minnesota	4	West Virginia	9
Mississippi	20	Wisconsin	36
Missouri	31	Wyoming	2

Drop Your Thesis!

Expeditions: 2009-2011
Leading Space Agency: ESA
Curriculum Grade Levels: Graduate (master's, Ph.D., M.D.)

Participating Countries: Austria, Belgium, Czech Republic, Denmark, Finland, France, Germany, Spain, Greece, Ireland, Italy, Luxemburg, Netherlands, Norway, Portugal, Romania, Sweden, Switzerland, United Kingdom

Cooperating Countries: Canada, Estonia, Hungary, Poland, Slovenia

Number of Graduate Students (master's, Ph.D., M.D.): 9 (period 2009–2011)
Number of Academic Supervisors: At least one per each student team
Number of Universities: 3
Ongoing Possibilities: Drop Your Thesis! is an annual opportunity; normally one team is accepted per year.

Project Description: Each year, teams composed of master's or Ph.D. students can apply to the Drop your Thesis! program by submitting an experiment proposal to the ESA Education Office. A review board composed of experts from the European Low Gravity Research Association (ELGRA), the ESA Directorate of Human Space Flight, ZARM-FABmbH and the ESA Education Office selects the best proposed microgravity experiment and gives the student team the opportunity to perform a series of launches using the ZARM drop tower in Bremen, Germany. During this Drop Your Thesis! campaign, the experiment payload is integrated into a drop capsule, which is allowed to fall freely, attaining microgravity levels as low as 10^{-6} g. The ZARM facility features a 146-meter-high tower and offers two drop modes: standard, in which the drop capsule is released from a height of 120 meters, and catapult, in which the capsule is propelled vertically to the top of the tower and afterward drops back down. The resulting experiment microgravity periods are 4.74 seconds and 9.3 seconds, respectively.

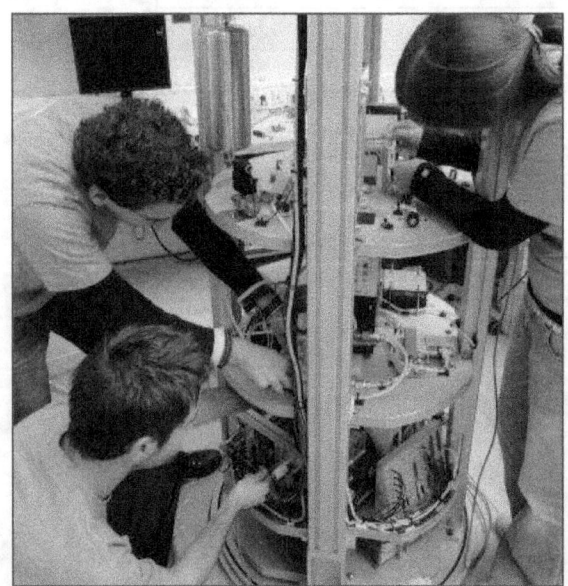

Hands-on experience for the Bubjet team. Image courtesy of ESA.

Education Lead: Piero Galeone, European Space Research and Technology Centre (ESTEC), European Space Agency, Netherlands

Website: http://www.esa.int/SPECIALS/Education/SEM67TLX82G_0.html

Fly Your Thesis!

Expeditions: 2009–2012

Leading Space Agency: ESA

Curriculum Grade Levels: Graduate (master's, Ph.D., M.D.)

Participating Countries: Austria, Belgium, Czech Republic, Denmark, Finland, France, Germany, Spain, Greece, Ireland, Italy, Luxembourg, Netherlands, Norway, Portugal, Romania, Sweden, Switzerland, United Kingdom

Cooperating Countries: Canada, Estonia, Hungary, Poland, Slovenia

> *Quentin Galand in the 2011 campaign: "All of us have profited and learned from many different things: project management, technical buildup of the experimental setup, the scientific objectives of the experiment and, last but not least, the great human experience."*

Number of Students: College (undergraduate) 30 (period 2009–2011)

Number of Academic Supervisors: At least one per student team

Number of Universities: 13

Ongoing Possibilities: A Fly Your Thesis! campaign is being organized in 2012 (with 14 students from four different universities). In 2013, the Fly Your Thesis! program will be suspended until further notice.

Description of Student Activity: The Fly Your Thesis! program gives university students the opportunity to fly their scientific experiments in microgravity as part of their master's thesis, Ph.D. thesis or research program by participating in a series of parabolic flights. These parabolic flights take place on the Airbus A300 Zero-G, which is operated and maintained by Novespace in Bordeaux, France.

Parabolic flights were proposed by ESA to foster university student interest in microgravity research all over Europe. The Student Parabolic Flight Campaign started in 1994 and was the first hands-on activity offered by ESA. The last full ESA student campaign was held in 2006. Thereafter, the program was thoroughly reviewed, and several managerial and safety recommendations were made for future campaigns. Following these recommendations and interactions between the ESA Education Office and the ESA Directorate of Human Space Flight, a new program called Fly Your Thesis! – An Astronaut Experience was defined. Teams composed of master's or Ph.D. students can apply to the Fly your Thesis! program by submitting a letter of intent to the ESA Education Office. A review board composed of experts from ELGRA, the ESA Directorate of Human Space Flight, and the ESA Education Office preselects as many as 20 teams and invites them to develop more detailed scientific and technical proposals. To conclude this pre-selection phase, the student teams present their projects to the review board during a dedicated workshop held at an ESA center. After this workshop, as many as four teams are selected to build and perform their experiment during an ESA parabolic flight campaign. The students accompany their experiments on board the A300 Zero-G aircraft for a series of three flights of 30 parabolas each, with each parabola providing approximately 20 seconds of microgravity. During this campaign, students work in close contact with several professional teams composed of renowned European scientists carrying out their own microgravity research.

Education Lead: Piero Galeone, ESTEC, ESA, Netherlands
Website: http://www.esa.int/SPECIALS/Fly_Your_Thesis/

Students from the University of Amsterdam investigate the water repellence of topsoil in the ARID experiment. Image courtesy of ESA.

NanoRacks NanoLabs

Expeditions: 25/26, 29/30
Leading Space Agency: NASA
Curriculum Grade Levels: 9–12 (secondary)
Impact: The first ISS Project Team from Valley Christian High School consisted of 22 students. Four Valley Christian High School ISS student teams consisting of 38 students were formed the next year; three ISS Project Partner high schools — Fremont Christian, Faith Christian Academy and Whittier Christian — had a total of 45 students.

Participating Country: United States
Participating State: California

Number of 9–12 Students (secondary): 105
Number of Schools: 4
Number of Teachers: 7

Brief Research Summary: Expeditions 25/26 NanoRacks NanoLabs include the following experiments:

- **NanoRacks-Valley Christian High School-Plant Seed Growth (NanoRacks-VCHS-Plant Seed)** contains an extended-sized autonomous, self-contained plant-seed growth experiment that is plugged into a NanoRacks Platform on board the ISS. The internal camera provides snapshots of the stages of growth during the experiment.

Expeditions 29/30 NanoRacks NanoLabs include the following experiments:

- **NanoRacks-Faith Christian Academy-Concrete Mixing Experiment (NanoRacks-FCA-Concrete Mixing)** compares the strength and molecular structure of concrete mixed by two different methods in microgravity with those of similar ground-mixed concrete. The comparison will be made using an atomic force microscope when the NanoLab is returned to Earth.

- **NanoRacks-Fremont Christian High School-Micro-Robot (NanoRacks-FCHS-Robot)** examines the effects of microgravity on remotely controlled robot control mechanisms and mechanical devices. The students programmed the fan-propelled robot to move in an x/y plane. The rate of travel and the total amount of travel are measured by analyzing photos.

- **NanoRacks-Valley Christian High School-*Bacillus Subtilis* Bacteria Growth (NanoRacks-VCHS-*B. Subtilis*)** studies the growth and growth rate of *Bacillus subtilis* bacteria in microgravity. Beef broth is added to the *Bacillus subtillis,* and the rate of growth is determined by measuring the amount of light shining through the bacteria-broth solution over time by using lumen sensors and analyzing the bacteria photos.

- **NanoRacks-Valley Christian High School-Electromagnetic Effects on Ferrofluid (NanoRacks-VCHS-Electromagnetic Ferrofluid)** studies the effects of a variable magnetic field on ferrofluids in microgravity.

Three-dimensional figures formed by the varying magnetic field are photographed and downlinked to the students daily to be analyzed and compared with those formed on the ground.

- **NanoRacks-Valley Christian High School-Electroplating (NanoRacks-VCHS-Electroplating)** studies the effects of electroplating gold and bronze in microgravity aboard the station and compares the results of identical electroplating on Earth. Samples of electroplating in space and on the ground are analyzed with Valley Christian's atomic force microscope.

- **NanoRacks-Valley Christian High School-Plant Growth (NanoRacks-VCHS-Plant Growth)** examines the growth and growth rate of marigold and thyme seeds in microgravity. With the bank of light-emitting diodes on to simulate the sun, two Wisconsin fast plants and two English thyme plants are watered at a predetermined rate and photographed to measure the rate of growth. The growth is compared with the growth of the same plants grown on the ground.

- **NanoRacks-Whittier Christian High School-*E.Coli* Bacteria and Kanamycin Antibiotic (NanoRacks-WCHS-*E. Coli* and Kanamycin)** studies the growth of green fluorescent protein tagged *Escherichia coli* (*E. coli*) bacteria in microgravity and the *E. coli* bacteria's resistance to the antibiotic Kanamycin by varying the antibiotic dosage. Photos of the bacteria are taken to determine its growth as measured by the amount of fluorescence. The bacteria's resistance to antibiotics is determined by measuring the decrease in fluorescence.

Description of Student Participation and Activities: The first ISS project team from Valley Christian High School consisted of 22 students. Four new Valley Christian High School International Space Station student teams were formed consisting of 38 Valley Christian High School students. Three other International Space Station Project Partner high schools — Fremont Christian, Faith Christian Academy and Whittier Christian — had a total of 45 students, and each school formed one team. Each student team was given a NanoLab to house its microgravity science experiments. Each one of the seven teams was responsible for selecting, designing, building, programming, testing and delivering its NanoLab to the Valley Christian High School International Space Station Student Integration Team.

Principal Investigators: Valley Christian High School, San Jose, California; Faith Christian Academy, Coalinga, California; Whittier Christian High School, La Habra, California; and Fremont Christian High School, Fremont, California, USA

Education Website: http://www.vcs.net/mathscience/iss-project/2011-12/index.aspx

NanoRacks-National Center for Earth and Space Science Education-1, -2, and -Aquarius (NanoRacks-NCESSE-1, -2, and -Aquarius)

Expeditions: 27/28, 31/32, 33/34

Leading Space Agency: NASA

Curriculum Grade Levels: K–8 (elementary), 9–12 (secondary), postsecondary

Impact: As a result of the NCESSE Student Spaceflight Experiments Program (SSEP) announcements of opportunity, 39 communities across the United States have joined the program, providing 71,900 students in grades 5 through 14 the opportunity to participate.

Participating Country: United States

Participating States: Arizona, California, Connecticut, Florida, Iowa, Illinois, Indiana, Kentucky, Louisiana, Massachusetts, Maryland, North Carolina, Nebraska, New Jersey, New Mexico, New York, Ohio, Oregon, Texas, Utah, Washington and the District of Columbia

Number of 5–14 Students (elementary/secondary/postsecondary): 71,900

Number of Schools: 193

Number of Teachers: 1756

Ongoing Opportunities: About 50 schools per year

Brief Research Summary: NanoRacks-NCESSE-1, -2 and -Aquarius are part of a commercial program that incorporates the science projects of schools from across the United States. Students design their own experiments using flight-approved fluids and materials that are flown on the Materials Diffusion Apparatus (MDA) or MixStix NanoLabs in a NanoRacks module. Students complete proposals for a flight opportunity, experience a science proposal review process, complete a flight safety review and attend their own science conference. The goal of this program is to allow students to experience scientific exploration through their own involvement.

Description of Student Participation and Activities: The SSEP, launched by the NCESSE in partnership with NanoRacks, LLC, is a remarkable commercial U.S. STEM education initiative that provides middle and high school students (grades 5–12) and undergraduates at two-year community colleges (grades 13–14) the ability to design and propose experiments to fly in low Earth orbit aboard the final flights of the space shuttle or the Soyuz and then aboard the International Space Station.

Principal Investigator: Jeff Goldstein, Ph.D., NCESSE, Capitol Heights, Maryland, USA

Education Website: http://ncesse.org/

Table 5 – NanoRacks-NCESSE-Aquarius Investigations

NanoRacks-NCESSE-Aquarius Investigation	School	Grade	City, State
Effect of Microgravity on the Antibacterial Resistance of *P. aeruginosa*	San Marino High School	10	San Marino, California
Microgravity Wine	Chaminade College Preparatory	9, 10	West Hills, California
How Does Parathyroid Hormone Affect Changes in Bone Mass in Microgravity?	Annie Fisher STEM Magnet School and University High School of Science and Engineering	8, 12	Hartford, Connecticut
Does Hay *Bacillus* Break Down Human Waste (Represented by Brown Egg) in Microgravity as Well as in Earth Gravity?	Stuart-Hobson Middle School	8	Washington, DC
Effect of Microgravity on Reproduction of Curli Producing *E. coli O157:H7 438950R*	Avicenna Academy	7	Lake County, Indiana
The Effect of Microgravity on the Quality and Nutritional Value of the Seed Sprout of a Germinated 92M72 Genetically Modified Soy Bean	Highland Christian School	7, 8	Lake County, Indiana
Killifish in Space	OA-BCIG High School	9–12	Ida County, Iowa
The Physiological Effects of Microgravity and Increased Levels of Radiation on Wild-Type and Genetically Engineered *Caenorhabditis elegans*	Henry E. Lackey High School	11	Charles County, Maryland
Effect of *Arthrobacter* on Polyethylene Decomposition Rate in Microgravity	Montachusett Regional Vocational Technical School	10, 12	Fitchburg, Maryland
Escherichia coli in Microgravity	Norris High School	12	Pleasanton and Norris, Nebraska
Spider Development and Gravity	Quebec Heights	5	Cincinnati, Ohio
Yeast in Space!	Cincinnati Gifted Academy	6	Cincinnati, Ohio
Hepatocyte Development in Bioscaffolds infused with TGFB3 in Microgravity	Johnston Middle School	8	Houston, Texas
Will Vitamin C Preserve Bone Density in Microgravity?	Parker Elementary School	5	Houston, Texas
The Effect of Microgravity on the Use of Cactus Mucilage for Water Purification	El Paso Community College Valle Verde Campus	14	El Paso, Texas

Table 6 – NanoRacks-NCESSE-2 Investigations

NanoRacks-NCESSE-2 Investigation	School	Grade	City, State
Microgravity Yeast Experiment	Parkridge Elementary School	7	Peoria, Arizona
Microgravity's Effect on Tomato Growth	Annie Fisher STEM Magnet School	8	Hartford, Connecticut
Will Microgravity Effect the Development of Goldfish?	Skinner West Classical, Fine Arts, and Technology School	5	Chicago, Illinois
All Mixed Up (Based on Gause's 1932 Experiment): The Effect of Microgravity on the Interaction of *Paramecium bursaria* and *Paramecium caudatum* in a Mixed Culture, Using Yeast and Bacteria as a Food Source	Avicenna Academy and Life Learning Cooperative	4–6, 4–12	Crown Point, Indiana
How Does Microgravity Affect the Maximum Cell Size of Tardigrades?	Ridge View High School	9–11	Galva-Holstein, Iowa
Physiological effects of microgravity on germination and growth of *Arabidopsis thaliana*	Henry E. Lackey High School	9–12	Charles County, Maryland
The Growth Rate of *Lactobacillus acidophilus* in Microgravity	Montachusett Regional Vocational Technical High School	11	Fitchburg, Massachusetts
Effects of Microgravity on Goodstreak Wheat	Potter-Dix Schools	6–12	Potter and Dix, Nebraska
The Effects of Microgravity on Oil Production in Salt-Stressed *Chlamydomonas reinhardtii*	Lincoln Public Schools Science Focus Program	11–12	Lincoln, Nebraska
Effects of Microgravity on Osteoblast Specialization and Bone Growth	Bridgewater-Raritan High School	11–12	Bridgewater-Raritan, New Jersey
Deposition and Formation of Zinc Phosphate Crystals in Microgravity	Yeshiva Ketana of Long Island	6–7	Inwood, New York

Table 7 – NanoRacks-NCESSE-1 Investigations

NanoRacks-NCESSE-1 Investigation	School	Grade	City, State
Development of Prokaryotic Cell Walls in Microgravity	Shelton High School	12	Shelton, Connecticut
Apples in Space	Crystal Lake Middle School	8	Broward County, Florida
The Effect of Microgravity on the Ability of Ethanol to Kill *E. Coli*	Maitland Middle School	8	Orange County, Florida
Efficiency of Microencapsulation in Microgravity as Compared to Gravity	Lincoln Hall Middle School	6	Lincolnwood, Illinois
The Effect of Microgravity on the Viability of *Lactobacillus GG*	The Academy@Shawnee	9–11	Jefferson County, Kentucky
What is the Effect of Microgravity on the Growth Rate of Murine Myoblasts?	Copper Mill Elementary School	5	Zachary, Louisiana
Swimming Patterns and Development of Zebra Fish After Exposure to Microgravity	Esperanza Middle School	8	Saint Mary's County, Maryland
Honey as a Preservative on Long-Duration Space Flights	Harry A. Burke High School	10	Omaha, Nebraska
Effects of Microgravity on Lysozyme's Antibacterial Properties	Omaha North High Magnet School	12	Omaha, Nebraska
Does the Radiation Exposure Effect Seed Germination Without the Protection of the Ozone Layer?	Tse' Bit'Ai Middle School	8	Shiprock, New Mexico
The Development of Minnow Fish Eggs in Space	Milton Terrace South Elementary School	5	Ballston Spa, New York

Reduced-Gravity Education Flight Program

Expeditions: Ongoing

Leading Space Agency: NASA

Curriculum Grade Levels: K–8 (elementary), 9–12 (secondary), College (undergraduate)

Impact: 3389 student flyers (does not include ground crew), 212 institutions, 696 teams, 49 states (plus the District of Columbia and Puerto Rico).

Participating Country: United States

Participating States: Alabama, Alaska, Arizona, Arkanasa, California, Colorado, Connecticut, Florida, Georgia, Hawaii, Idaho, Illinois, Indiana, Iowa, Kansas, Kentucky, Maine, Maryland, Massachusetts, Michigan, Minnesota, Mississippi, Missouri, Montana, Nebraska, Nevada, New Hampshire, New Jersey, New York, North Carolina, North Dakota, Ohio, Oklahoma, Oregon, Pennsylvania, Rhode Island, South Carolina, South Dakota, Tennessee, Texas, Utah, Virginia, Washington, West Virginia, Wisconsin, Wyoming, plus the District of Columbia and Puerto Rico

Number of Students (elementary): 3389

Number of Schools: 212

> *"The program is definitely the kind of thing NASA needs more of because in reality all these kids who dream of being astronauts probably won't be selected. I read a statistic that the chances of being an astronaut were slimmer than being a professional athlete who gets struck by lightning! So really this is a program that lets kids get as close as they can to their dreams."*

Experiment Description: Before experiments are sent to the station, they are tested using a variety of microgravity simulations. Many new pieces of experimental hardware and research approaches are tested in parabolic flight simulations of microgravity. During these parabolic flights, experimenters have access to repeated short periods (25 seconds) of microgravity in which to conduct their experiments. Many of the microgravity simulations in parabolic flight are directly or indirectly linked to ISS research.

> *"Hey, I did something awesome and it involved science and experimentation and engineering. I really think this program does a great job of doing that and is worth every second of the work — even without the flight. You learn how to actually put together real research and prepare a real experiment to fly. That kind of experience is invaluable and can only serve as a benefit in my future endeavors."*

Description of Student Participation and Activity: The Reduced Gravity Education Flight Program consists of two areas that allow for student participation. The first area is that of research prior to an experiment being sent to ISS. Students who are affiliated with a principal investigator have assisted in many capacities. These students design, build and test hardware as well as collect and analyze data from the microgravity simulation flights. The overall experience includes scientific research, hands-on experimental design, test operations and educational/public outreach activities. The objectives of the program are (1) to provide students and educators with an outstanding educational opportunity to explore microgravity; (2) to attract outstanding young scholars to careers in math, science and engineering in general; (3) to introduce young scholars to careers with NASA and in the space program in particular; (4) to provide a platform for students and educators to understand how microgravity affects research and testing of serious science and engineering ideas; and (5) to provide an opportunity for both the general public and school children to discover educational and professional opportunities available at NASA.

Description of Teacher Participation and Activities: This program is also open to K–12 teachers across the country. More information about program specifics and annual reports can be found on the following website: http://microgravityuniversity.jsc.nasa.gov/theArchives/annualreports.cfm.

Education Leads: Douglas Goforth and Sara Malloy, NASA JSC, Houston, Texas, USA
Education Website: http://microgravityuniversity.jsc.nasa.gov

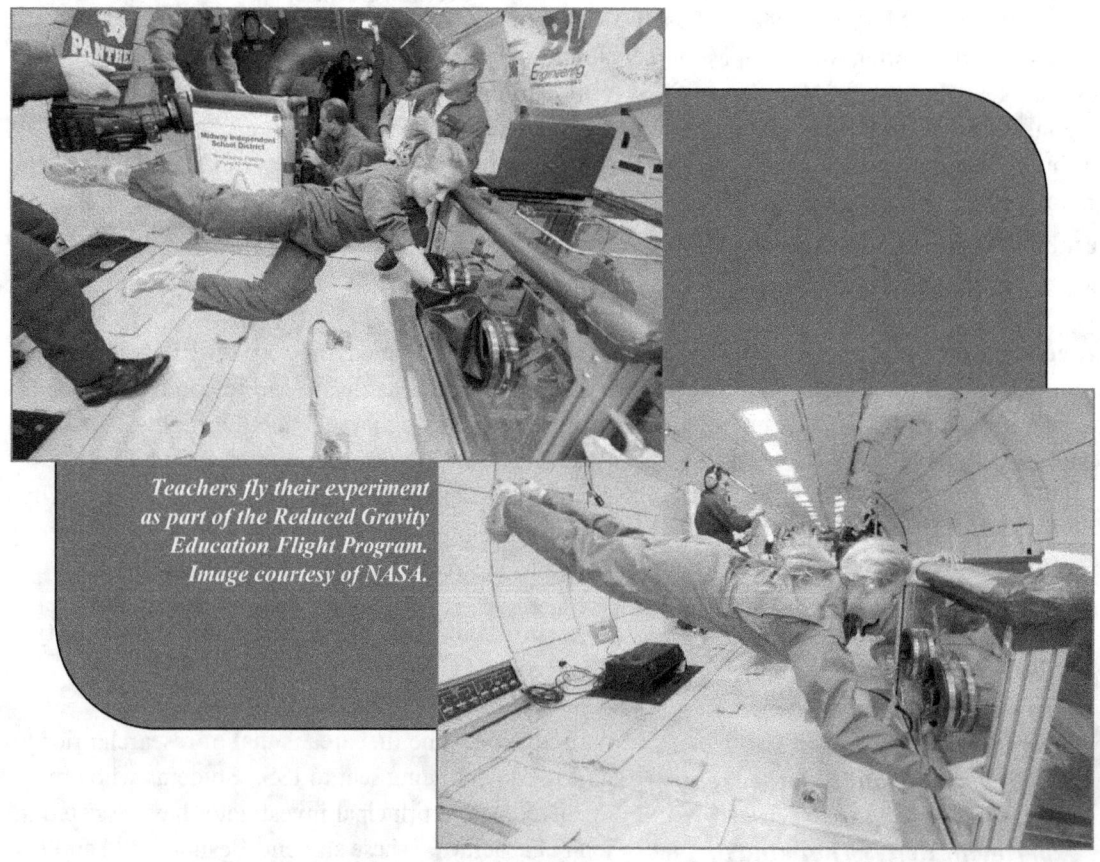

Teachers fly their experiment as part of the Reduced Gravity Education Flight Program. Image courtesy of NASA.

Rocket and Balloon Experiments for University Students (REXUS/BEXUS)

Expeditions: Ongoing

Leading Space Agencies: German Aerospace Center (DLR) and the Swedish National Space Board (SNSB)

Participating Space Agency: ESA

Participating Countries: Austria, Belgium, Czech Republic, Denmark, Finland, France, Germany, Spain, Greece, Ireland, Italy, Luxemburg, Netherlands, Norway, Portugal, Romania, Sweden, Switzerland, United Kingdom

Cooperating Countries: Canada, Estonia, Hungary, Poland, Slovenia

Number of University Students: 146 (period 2008–2011)

Number of Academic Supervisors: At least one per student team

Number of Universities: 31

Ongoing Possibilities: REXUS and BEXUS are yearly opportunities

Description of Student Activity: The REXUS and BEXUS programs offer opportunities for student experiments to be flown on sounding rockets and stratospheric balloons. Each flight carries a payload consisting solely of student experiments.

The REXUS/BEXUS programs are realized under a bilateral agency agreement between the DLR and the SNSB. The Swedish share of the payload has been made available to students from other European countries through collaboration with the ESA.

EuroLaunch, a collaboration between the Esrange Space Center of the Swedish Space Corporation (SSC) and the Mobile Rocket Base (MORABA) of DLR, is responsible for the campaign management and operations of the launch vehicles. Experts from ESA, SSC and DLR provide technical support to the student teams throughout the project.

REXUS is an unguided, spin-stabilized, solid-propellant, single-stage rocket. The vehicle has a length of 5.6 meters and a diameter of 35.6 centimeters, and the total available mass for student experiments is about 30 kilograms.

The BEXUS balloon has a volume of 12,000 cubic meters and a diameter of 14 meters when filled. The total mass available for the experiments is between 40 and 100 kilograms. The maximum altitude is 35 kilometers, and the flight duration is two to five hours.

Education Lead: Piero Galeone, ESTEC, Netherlands

Websites: http://www.esa.int/SPECIALS/Education/SEMTTQJV3AF_0.html, http://www.rexusbexus.net/

JAXA Seeds in Space I (Asagao and Miyako-Gusa)

Expeditions: Ongoing

Leading Space Agency: JAXA

Curriculum Grade Levels: K–8 (elementary), 9–12 (secondary)

Participating Country: Japan

Number of K–8 Students (elementary): 5994

Number of 9–12 Students (secondary): 2716

Number of Schools: 201

Number of Teachers: 201

Description of Student Participation and Activities: In March 2008, seeds of Asagao (Japanese morning glory) and Miyako-gusa (Japanese bird's foot trefoil) were launched to the International Space Station aboard space shuttle Endeavour, STS-123. The specimens remained on the ISS for nine months and were returned to Earth. In May 2010, the space-flight seeds of Asagao and Miyako-gusa were distributed to hundreds of schools in Japan, with another set of ground (negative) control (never left Earth) and positive control (irradiated with carbon ion beams at RIKEN Accelerator Research Facility) seeds. Each group of participating students, 3 to 18 years old, was given 10 flight, ground control and positive control seeds as one set for Asagao; for Miyako-gusa, 20 seeds were distributed. The students continued to cultivate their plants twice for two seasons, until the fall of 2011, to find the mutants.

Description of Teacher Participation and Activities: Teachers prepare for student activities associated with this experiment with a set of seeds. Several educator guides involving cultivation methods for Asagao or Miyako-gusa, mutant identification methods for Asagao or Miyako-gusa, past examples of Seeds in Space education experiments, cosmic radiation, the role of higher plant mutants to agriculture and their biological safety have been distributed to each teacher who supervises students in classrooms. Furthermore, for each month's cultivation period, teachers will review their own students' crude data and publish the data on the JAXA home page. The summation of all the school data becomes one experiment. After two rounds of cultivation, teachers will send the final cultivation reports to JAXA, which will perform a final assessment of mutant occurrences.

Education Lead: Tamotsu Nakano, Ph.D.

Education Website: Japanese: http://edu.jaxa.jp/seeds/
English: http://edu.jaxa.jp/en/education/international/ISS/SIS/

> *Ms. Yoko Akamine, a teacher of Oitaminami High School, Oita City, Oita Prefecture: "Since we think this is a very interesting and rare experiment, we would like to tackle it as one of our Science Club activities."*

Additional Comments on Student/Teacher Involvement in the Investigation: It is JAXA's hope that the students understand that the mutation rates of seeds in space are extremely low compared to their expectations and that they learn what scientific experiments are. The final goal of the Seeds in Space education program is to have students experience the science and participate in a real scientific investigation.

Students scatter seeds of Asagao (Nara University of Education Junior High School, Nara City, Nara Prefecture, Japan). Image courtesy of Nara University of Education Junior High School and JAXA.

Space flight Miyako-gusa, pictured to compare three experimental groups (Wakayama Shin-ai Girls' Junior and Senior High School, Wakayama City, Wakayama Prefecture, Japan). Image courtesy of Wakayama Shin-ai Girls' Junior and Senior High School and JAXA.

JAXA Spaceflight Seeds Kids I (Himawari)

Expeditions: Ongoing

Leading Space Agency: JAXA

Curriculum Grade Levels: K–8 (elementary)

Participating Country: Japan

Number of K–8 Students (elementary): 2054

Number of Schools: 44

Number of Teachers: 44

Description of Student Participation and Activities: Schools participating in this program were provided the opportunity to grow sunflower seeds that flew on the ISS for approximately nine months in 2008 as a start to Touch the Universe. Through a series of education guides, a deeper understanding of the past and near-term future of human space flight, as well as that of Earth and life, will be learned. In fact, JAXA intended to have school children experience the pseudo overview effect by learning about harsh space circumstances disclosed by recent space developments. As a result, it would be much easier for them to find that Earth is really not only an oasis of space but also a cradle for living things on the Earth. Also, a contest was held for scientific reporting on lunar outpost agriculture by campaign participants. All contestants received an hourglass gift filled with simulated lunar regolith (※). (Note: The hourglass does not reflect time accurately.)

Space flight Himawari, pictured by students (Tateyama-Sazanami School, Tateyama City, Chiba Prefecture, Japan). Image courtesy of Tateyama-Sazanami School and JAXA.

Description of Teacher Participation and Activities: To prepare for student activities associated with this experiment, several educator guides involving sunflower cultivation, the moon's surface (including lunar regolith), soil and soil organisms, mycorrhizal fungi, space farms in microgravity and extraterrestrial agriculture (lunar and Martian agriculture) were mailed to each teacher who supervises students in classrooms. After seed cultivation, teachers led the students in writing the final scientific report on lunar outpost agriculture, which was primarily proposed by students, to take part in JAXA's contest.

Additional Comments on Student/Teacher Involvement in the Investigation: Sponsored by the Space Education Center of JAXA, which provides opportunities to use the International Space Station as a new field of education, this is the first campaign of JAXA Spaceflight Seeds Kids, designed primarily to promote interest in science and inspire school children.

Education Lead: Tamotsu Nakano, Ph.D.
Education Website: Japanese: http://edu.jaxa.jp/himawari/
English: http://edu.jaxa.jp/en/education/international/ISS/SSK/

> *Mr. Kanta Honma, Teacher of Higashikagura Shibinai Elementary School, Higashikagura Town, Hokkaido:* "It is a small-scale elementary school of just 13 children, located at the foot of Asahi-dake, the highest peak of Hokkaido. At night, we always have a lot of stars in the heaven. I would like to make them begin to interest themselves in space and space development, taking advantage of the cultivation of the space-flight sunflower."
>
> *Miss Miku Fujii, third grade student of Ginganosato Elementary school, Fukuyama City, Hiroshima Prefecture:* "When I played a class helper of the day, I just took photographs of sunflowers. We have taken care of sunflowers, which were growing on and on, day by day. We really enjoyed every day in this season. Thank you, JAXA."

Take Your Classroom Into Space (TYCIS)

Expeditions: 19/20, 21/22, 27/28, 29/30, 31/32
Leading Space Agency: ESA
Participating Space Agencies: Italian Space Agency (ASI), Netherlands Space Office (NSO)
Impact: Thousands of children and hundreds of schools throughout Europe have participated in Take Your Classroom Into Space activities, either during live in-flight calls or in their classrooms after completion of the expedition.

Participating Countries: Austria, Belgium, Czech Republic, Denmark, Finland, France, Germany, Spain, Greece, Ireland, Italy, Luxemburg, Netherlands, Norway, Portugal, Romania, Sweden, Switzerland, United Kingdom

Number of Primary School Students: 10,500
Number of Secondary School Students: 10,500
Number of Teachers: 1300
Number of Schools: 1300
Ongoing Possibilities: TYCIS activities are ongoing, with a different activity for each ESA astronaut.

Description of Student Activity: The TYCIS activity was initially developed in 2009 when Frank de Winne started his OasISS mission, the first long-term ESA astronaut mission. The principle of TYCIS is simple: develop an educational payload destined to fly on the International Space Station. A crewmember performs the experiment either live during an in-flight call or records it and downlinks it at a later time. Schools that wish to participate in TYCIS order the free TYCIS kits, which resemble the flight hardware.

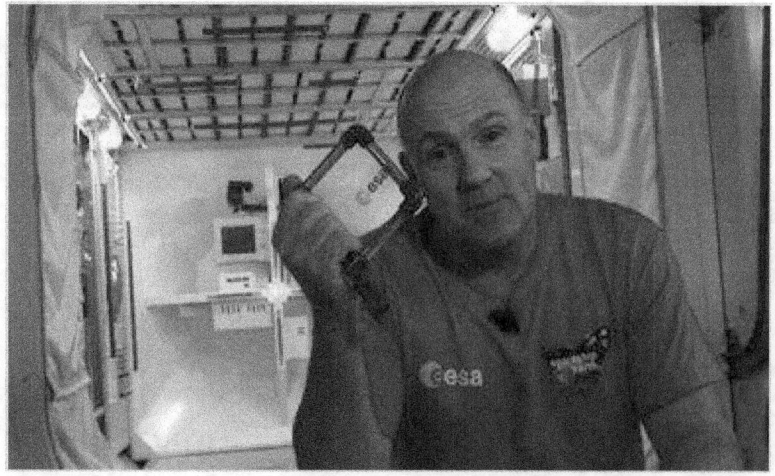

ESA astronaut André Kuipers with an Educational Payloads Operations (EPO) Convection experiment. Image courtesy of ESA.

The objective is for the students to perform the control experiment on Earth, bound by Earth's gravity, while the astronaut performs the same experiment in weightlessness. The differences in outcome highlight the influence of gravity on the phenomenon being investigated. The TYCIS is accompanied by lessons translated into several languages; the lessons can be accessed at any time via the ESA education website, even after completion of the expedition. Similarly, the educational kits also can be ordered while stocks last.

The TYCIS activity also incorporates an educational in-flight call to the astronaut who is performing or has performed the relevant TYCIS activity. ESA enables up to four sites (museums or science centers) in Europe to interact with the astronaut directly by asking questions. These events typically involve 1300 children per in-flight call. The in-flight calls are recorded and available on ESA's Web streaming network (wsn.spaceflight.esa.int).

Education Lead: Nigel Savage, Ph.D., ESTEC, Netherlands
Website: http://wsn.spaceflight.esa.int/education

ESA astronaut André Kuipers with an EPO foam stability experiment.
Image courtesy of ESA.

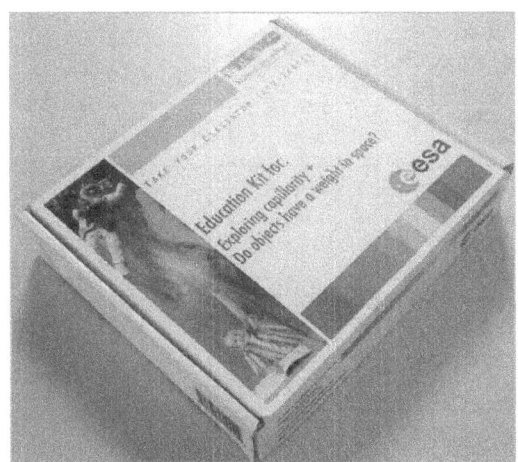

The 2009 TYCIS kit for a capillarity and mass measurement experiment. Image courtesy of ESA.

Table 8 – ESA Education Payload Operations and Demonstrations

Demonstration	Objective	Expedition	Crewmember
EPO TYCIS – Capillarity	This experiment, by Belgian teacher Jef Luyten and his Italian colleague Luigi Lombardo, was about capillarity. Here on Earth, the capillary effect can be seen in fine tubes containing liquid: surface tension pulls the liquid column up until there is a sufficient mass of liquid for gravity to overcome the intermolecular forces. As the mass of the liquid column is proportional to the square of the tube's diameter, a narrow tube will draw a liquid column higher than a wide tube.	19/20 21/22	Frank DeWinne
EPO TYCIS – Mass Measurement	This experiment, proposed by Theodoros Pierratos of Greece, Anicet Cosialls from Spain and Mieke Recour in Belgium, looked at calculating the mass of an object in space by timing its oscillations on a spring. It was conducted in a live link from the ISS September 2009. The experiment clearly illustrated the difference between the concepts of weight and mass.	19/20 21/22	Frank DeWinne
EPO TYCIS – Greenhouse in Space	This experiment attempted to yield seeds from plants grown from seed. The greenhouse, especially designed for growing plants in microgravity, was watered initially during a live link with Paolo Nespoli in February 2011. Hundreds of school children also initiated their experiments at the same time and carefully monitored the plant growth in the following weeks. Unfortunately, the experiment on board the ISS was terminated early due to growth of mold in the specially designed greenhouse.	25/26	Paolo Nespoli
EPO TYCIS – Convection	This demonstration illustrates to students on a small scale how thermal gradients drive convective currents and, on the large scale of a planet, how temperature gradients influence density-driven convection and create atmospheric and oceanic currents.	29/30, 31/32	André Kuipers
EPO TYCIS – Foam Stability	This experiment demonstrates physical properties of wet foams and how gravity influences stability as well as how understanding foam formation and stability can help us make cutting-edge materials. André Kuipers shows students how a foam is created from pure water in microgravity and observes its stability.	29/30, 31/32	André Kuipers

Tomatosphere-III

Expeditions: 19/20, 27/28-31/32, Ongoing

Leading Space Agency: CSA

Curriculum Grade Levels: 3–8 (elementary), 9–10 (secondary)

Impact: More than 13,000 classrooms are growing seeds, comparing the growth to space-flown and control seeds.

Participating Country: Canada

Number of Schools: 12,400

Brief Research Summary: Tomatosphere-III is an educational investigation that uses the excitement of space exploration as a medium for teaching students about science, space, agriculture and Canada's role in support of long-term space flight. The Tomatosphere program began in 2001 and has expanded each year; the investigation now involves more than 13,000 classrooms in Canada. In the spring of each year, classrooms conduct experiments to investigate the effects of the space environment on the growth of tomato seeds in support of long-duration human exploration, which includes establishing a base on the moon and later proceeding to Mars. The students' findings will begin to address the question of how we supply long-duration space exploration missions with the life support requirements of food, water and oxygen. Currently, space vehicles are able to carry just enough of these to service the crew for short missions; supplies for long-duration crews on the International Space Station are currently refreshed by visiting space vehicles.

> *Larry Burroughs, Cheslakees Elementary School, British Columbia:* "The students really enjoyed this project and related very well to the project. They made a real connection to Canada's space program and thought it was very exciting that they were working with seeds that had travelled into outer space! I am most appreciative of all the sponsors for their part in providing a relevant, real-life classroom experience for the students in my school."

> *Jo-Ann LaCharity, Castor Valley Elementary School, Ontario:* "These authentic experiences are essential for hooking children on science and provide a wealth of learning opportunities."

Description of Student Participation and Activities: Students will compare the germination rates of the control group with that of the seeds exposed to the microgravity environment on board the ISS and will report on the growth and development of the plants. Students will learn how to conduct a scientific experiment and may be inspired to pursue further education in the areas of science and technology. It is essential that we focus on the technical needs of ventures like the space program to motivate young people to pursue studies in science and technology. Not all students can become astronauts; however, many will be able to find significant, worthwhile roles in providing support for knowledge-based programs such as those being developed by CSA and other agencies. Tomatosphere-III educational units include: (1) How to Feed a Martian (Grades 3–4): Nutrition Focus for Astronauts' trips to the Red Planet; (2) Surviving on the Red Planet (Grade 6): Recycling

Breathable Air; (3) The Martian Environment (Grades 7–8): Weather Station on Mars; (4) The Energy to Survive (Grades 9–10): Nutritional Requirements for Long-Duration Missions.

Principal Investigator: Jason Clement, CSA, Saint-Hubert, Quebec, Canada

Education Website: www.tomatosphere.org

During a previous Tomatosphere program, students studied the growth of their tomato plants in Miss Smith's grade three class at Langley Fundamental Elementary, Vancouver, British Columbia, Canada. The students took their plants home to grow in their gardens over the summer. Image courtesy of Tomatosphere.

Education Competitions

These educational activities involve a student design competition to which students submit proposals for experiments. The proposals are judged, and the students who win often have the opportunity to fly their designs on the International Space Station.

Kids in Micro-g

Expeditions: 23–28
Leading Space Agency: NASA
Curriculum Grade Levels: K–8 (elementary)

Participating Country: United States

Participating States: Alabama, California, Connecticut, Florida, Georgia, Idaho, Maryland, Michigan, Missouri, New Jersey, New York, Ohio, Oregon, Pennsylvania, South Carolina, Virginia, Washington

Number of K–8 Students (elementary): 951
Number of Schools: 50
Number of Teachers: 53

Description of Student Participation and Activities: Kids in Micro-g was an experiment design challenge for students in grades 5–8. Its purpose was to give students a hands-on opportunity to design an experiment or a simple demonstration that could be performed both in the classroom and aboard the International Space Station. The students were provided a list of materials to use in the classroom that can also be found on board the ISS. The winning experiments had observably different results when the experiments were performed in the 1-g environment of the classroom, compared to when the experiments were performed by astronauts in the micro-g environment of the ISS. Video of the astronauts performing the experiments was provided to the students to complete their 1 g versus micro-g investigations.

In 2011, the winning experiment, developed by a team of two fifth graders, was titled Attracting Water Drop, which determined if a free floating water drop can be attracted to a static-charged rubber exercise tube. What resulted was an exciting new understanding of physics in space.

The Kids in Micro-g project will be replaced by the YouTube Space Lab Competition in 2012.

Education Lead: Trinesha Dixon, NASA JSC, Houston, Texas, USA

Education Website:
http://www.nasa.gov/tfs

NASA astronaut Douglas (Doug) Wheelock, Expedition 25 Commander, performs the Liquids in Microgravity experiment from the Virginia Academy in Ashburn, Virginia.
Image courtesy of NASA.

NASA astronauts Catherine (Cady) Coleman and Ronald (Ron) Garan perform the Attracting Water Drops experiment from Chabad Hebrew Academy. Image courtesy of NASA.

Roger, Potlatch Elementary School, Potlatch, Idaho: "This experience will stick with me forever. This is very exciting to have our experiment be performed by trained astronauts."

Hanna, Hamlin School, San Francisco, California, as she views video of astronaut Doug Wheelock performing the experiment she designed: "It's kind of mind boggling. He's in space. He's floating. It's basically my idea in space. Well, it is my idea in space!"

Table 9 – Kids in Micro-g List of Winners

Kids in Micro-g Investigations	School	Grade	City, State
Water Absorption/Capillary: Determined the water absorption rates of two different materials.	Brownell Middle School	8	Grosse Pointe Farms, Michigan
Bottle Blowing in Space: Determined if blowing across the tops of bottles filled with different amounts of water will create the same tones in space as on Earth.	Vaughan Elementary www.cobbk12.org	5	Powder Springs, Georgia
Speed: Determined if the radius (of the circle of revolution) affects the speed at which an outer object travels around a central object, and whether microgravity will change the results in this experiment.	Hamlin School www.hamlin.org	7	San Francisco, California
Water and Hot Sauce: Determined if adding water to hot sauce in a microgravity environment will affect its surface tension.	Brownell Middle School	8	Grosse Pointe Farms, Michigan
Newton's Space Office: Tested Newton's Laws of Motion using a bag of paper clips.	East Hartford-Glastonbury Elementary Magnet School	5	East Hartford, Connecticut
Motion of Projectiles: Investigated the effects gravity has on the motion of slingshot projectiles. Speed, distance traveled and path of projectile specifically will be studied.	Carl Sandburg Middle School	6	Old Bridge, New Jersey
Low Gravity Artist: Studied human adaptability, focusing on the role that gravity plays in a human's ability to draw a picture.	Henry E. Lackey High School	8	Orlando, Florida
Liquids in Microgravity: Determined if liquid will move from its original position inside a bottle while in microgravity.	Virginia Academy	8	Ashburn, Virginia
Water Absorption: Tested the water absorption capabilities of various materials in a microgravity environment.	Vaughan Elementary	5	Powder Springs, Georgia

Table 10 – Kids in Micro-g-2 List of Winners

Kids in Micro-g-2 Investigations	School	Grade	City, State
Attracting Water Drops: This experiment determines if a free-floating water drop is attracted to a static charged rubber exercise tube.	Chabad Hebrew Academy www.chasd.org	5	San Diego, California
Flight of Paper Rockets Launched by Air Cannon: This experiment determines the direction and distance traveled by a paper air rocket launched in microgravity.	Neighborhood After School Science Association	5–8	Ava, New York
Pondering the Pendulum: This experiment examines the effects of microgravity on a pendulum.	Key Peninsula Middle School www.kpms.psd401.net	8	Lakebay, Washington
Pepper Oil Surprise: This experiment investigates the interaction of liquid pepper/oil and water in a plastic bag in microgravity.	Potlatch Elementary www.potlatchschools.org	6	Potlatch, Idaho
Buoyancy in Space: This experiment determines if the buoyancy of an object is affected in a microgravity environment.	Gate of Heaven School	8	Dallas, Pennsylvania
Dispersion of Liquid: This experiment compares the dispersal of liquid pepper in microgravity to its dispersal in Earth's gravity.	Will James Middle School in Billings www.williamsmiddleschool.org	6–7	Billings, Montana

Spaced Out Sports

Expeditions: 31/32, Ongoing

Leading Space Agency: NASA

Curriculum Grade Levels: 3–8 (elementary), 9–10 (secondary)

Participating Countries: Turkey, United States

Participating States: Alabama, Arizona, Arkansas, California, Connecticut, Indiana, Kentucky, Louisiana, Massachusetts, Minnesota, Mississippi, New Jersey, New York, Ohio, Pennsylvania, Vermont, Virginia,

Number of K–8 Students (elementary): 855

Number of Schools: 24

Number of Teachers: 24

Description of Student Participation and Activities: Spaced Out Sports is a national student design challenge geared toward grades 5–8. The purpose is for students to apply Newton's Laws of Motion by designing or redesigning a game for International Space Station astronauts to play in space. As students design a new sport, they learn Newton's Laws of Motion and the effect of gravity on an object. They learn to predict the difference between a game or activity played on Earth and one played in the microgravity environment of the ISS. Student teams submit game demonstrations via a playbook and a video; submissions are accepted from schools, home school groups, and after-school or enrichment programs.

Students involved in the Spaced Out Sports competition.
Image courtesy of Stennis Space Center.

Spaced Out Sports' winning teams are selected regionally and nationally by NASA Stennis Space Center's Education Office. The first place team is awarded a NASA school-wide or program-wide celebration. The top three teams' games are played on the ISS and recorded for a future broadcast, the date and time of which will be determined later. All contributing schools and programs participate in a Digital Learning Network (DLN) webcast with ISS NASA astronauts. The date and time of the webcast also will be determined later.

Education Lead: Cheryl Guilbeau, NASA Stennis Space Center, Mississippi, USA

Education Websites: http://www.nasa.gov/education/tfs,
http://education.ssc.nasa.gov/spacedoutsports.asp

Synchronized Position Hold, Engage, Reorient, Experimental Satellites (SPHERES) – Zero Robotics

Expeditions: 21/22-25/26, 29/30-33/34, Ongoing

Space Agencies: NASA, ESA

Curriculum Grade Levels: 6–8 (elementary), 9–12 (secondary), college (undergraduates)

Participating Countries: United States, European Union

Experiment Description: Zero Robotics is a student competition that takes arena robotics to new heights — literally. The robots are miniature satellites called SPHERES, and the final competition of every tournament is held aboard the International Space Station! Zero Robotics was created in 2009 by the Massachusetts Institute of Technology (MIT) Space Systems Laboratory (SSL) and astronaut Greg Chamitoff with the goal of opening research on the station to large groups of secondary school students. Zero Robotics draws significant inspiration from For Inspiration and Recognition of Science and Technology (FIRST) Robotics (hence the name) and the competitions share common goals, including building lifelong skills in science, technology, engineering and math. Zero Robotics is envisioned to be a competition complementary to FIRST, since FIRST Robotics concentrates on the building of hardware and human control techniques, while Zero Robotics concentrates on the development of autonomous software.

Gregory (Greg) Chamitoff, Ph.D., works with Synchronized Position Hold, Engage, Reorient, Experimental Satellites (SPHERES) in the U.S. Laboratory during Expedition 17. NASA image ISS017E015118.

In the fall of 2009, the SSL conducted a pilot program of the Zero Robotics competition with two schools from northern Idaho. The competition was motivated by the idea of a satellite assistant robot. The first robotics competition aboard the ISS took place December 9, 2009. Zero Robotics was a component of NASA's Summer of Innovation, a nationwide program targeted at encouraging STEM education for middle school students.

In the fall of 2010, Zero Robotics conducted a nationwide pilot tournament for high school students named the Zero Robotics SPHERES Challenge 2010: HelioSPHERES.

In 2011, ESA joined NASA and MIT in giving high school students all over Europe an opportunity to participate in the competition. Several schools from ESA member states are invited to create rival programs that control three SPHERES in real time on the space station.

Each European school is assigned a local SPHERES expert familiar with the coding requirements of the droids. Several university staff members were trained at MIT with ESA's sponsorship. The competition is not only about feeding the satellites sets of commands. The local experts help students build critical engineering skills, such as problem solving, design thought process, operations training and teamwork. Their results can lead to important advances for satellite servicing and vehicle assembly in orbit.

Teams in the U.S. and Europe test their algorithms under realistic microgravity conditions. They compete against each other in elimination rounds, with finals on both sides of the Atlantic.

The software of the top 10 winners is uploaded and run in the three weightless spheres by the astronauts. The exciting final is streamed live at ESA's technology center in the Netherlands, known as ESTEC, with a unique touch: a three-dimensional (3D) video is shown so that viewers can fully appreciate the guidance, navigation and control of the SPHERES.

The SPHERES – Zero Robotics program is led by MIT, TopCoder, and Aurora Flight Sciences under the sponsorship of Defense Advanced Research Projects Agency (DARPA) and NASA.

In 2012, three teams combined to win. The teams that made up Alliance Rocket were Team Rocket, River Hill High School, Clarksville, Maryland; Defending Champions, Storming Robots, Branchburg, New Jersey; and SPHEREZ of Influence, Rockledge High School, Brevard County, Florida. Alliance CyberAvo consisted of CyberAvo, I.T.I.S. Amedeo Avogrado, Turin, Italy; Ultima, Kaethe Kollwitz Oberschule, Berlin, Germany; and Lazy, Heinrich Hertz Gymnasium, Berlin, Germany.

Description of Student Participation and Activities: Students program the satellites to play a challenging game. Students can create, edit, share, save, simulate and submit code, all from a Web browser, to the Zero Robotics website. All tournaments are free of charge, and all that is required to participate is a team, mentorship and the Internet! An astronaut conducts the championship competition in microgravity with a live broadcast from the ISS.

The SPHERES – Zero Robotics Competition includes four steps in the process:

Step 1: Proposal Preparation and Submission

Students submit a proposal outlining their plans for the competition, specifically addressing algorithm development.

Step 2: Algorithm Development in Simulation

Students code and test the game algorithm in the MIT-developed simulation.

Students send final algorithm files to MIT for review and scoring.

Step 3: Ground Hardware Testing (Regional)

Students direct the testing of algorithm files on SPHERES satellites at multiple NASA and/or industry facilities that have flat floors around the country.

Step 4: International Space Station Hardware Testing

Students update algorithms based on lessons learned from ground hardware testing to be operational on the station SPHERES.

MIT packages, tests and sends the final files to NASA.

A crewmember runs the code in space in a test session (live video of session is sent to winning teams, which may be invited to MIT).

Principal Investigators: Jeffrey Hoffman and David W. Miller, MIT, Cambridge, Massachusetts; Nigel Savage, Ph.D., ESA

Education Websites: http://www.zerorobotics.org

YouTube Space Lab

Expeditions: 29/30, 31/32

Leading Space Agencies: NASA, ESA, JAXA, Global

Curriculum Grade Levels: 9–12 (secondary)

Participating Countries: Global

Participating States: Alabama, Alaska, Arizona, Arkansas, California, Colorado, Connecticut, Delaware, Florida, Georgia, Hawaii, Idaho, Illinois, Indiana, Iowa, Kansas, Kentucky, Louisiana, Maine, Maryland, Massachusetts, Michigan, Minnesota, Mississippi, Missouri, Montana, Nebraska, Nevada, New Hampshire, New Jersey, New Mexico, New York, North Carolina, North Dakota, Ohio, Oklahoma, Oregon, Pennsylvania, Rhode Island, South Carolina, South Dakota, Tennessee, Texas, Utah, Vermont, Virginia, Washington, West Virginia, Wisconsin, Wyoming

Number of 9–12 Students (secondary): 2,000 entries from 80 countries (list not available at time of this publication)

Description of Student Participation and Activities: YouTube and Lenovo, in cooperation with Space Adventures and space agencies, including NASA, ESA and JAXA, launched the YouTube Space Lab contest, a global initiative that challenges 14 to 18 year old students to design a science experiment that can be performed in space. The winning team's experiment is conducted aboard the International Space Station, making it the universe's largest science lesson streamed live for the world to see via YouTube. Space Lab is part of a larger YouTube for Schools initiative aimed at highlighting and providing educators access to the wealth of educational content available on YouTube and is also part of Lenovo's focus on equipping students with 21st century skills via personal computer (PC) technology.

Alone or in groups of up to three, 14 to 18 year-olds may submit a YouTube video describing their experiment to YouTube.com/SpaceLab. Of the 2,000 entries received from around the world, 60 finalists were selected. A prestigious panel of scientists, astronauts, and teachers judge the entries with input from the YouTube community. Six regional finalists gathered in the United States in March 2012 to experience a zero-gravity flight, and those who were not finalists received other prizes. The two global winners will see their experiments performed 250 miles into space and live streamed on YouTube in the summer. These experiments examine the predatory behavior of a jumping spider and the anti-fungal properties of *Bacillus subtilis*, a naturally occurring bacteria that is commonly used as an anti-fungal agent for agricultural crops. Also, the global winners will choose a unique space experience as a prize: either a trip to Japan to watch their experiment blast off to space or a trip to Russia for an authentic space training experience at the facilities where Yuri Gagarin became a cosmonaut.

YouTube Space Lab is one component of YouTube for Schools, a new opt-in program that allows educators to access YouTube's broad library of educational content from inside their school network. Through the YouTube for Schools network setting, schools can access the educational videos housed under YouTube.com/EDU while filtering out non-educational content. They may also select unlimited additional channels not included under YouTube.com/EDU to make them accessible inside their school network (visit YouTube.com/teachers for more information).

Principal Investigator: Zahaan Bharmal, Google, London, UK; Dorothy Chen, Troy, Michigan, USA; Sara Ma, Troy, Michigan, USA; Amr Mohamed, Alexandria, Egypt

Education Website: YouTube.com/Spacelab

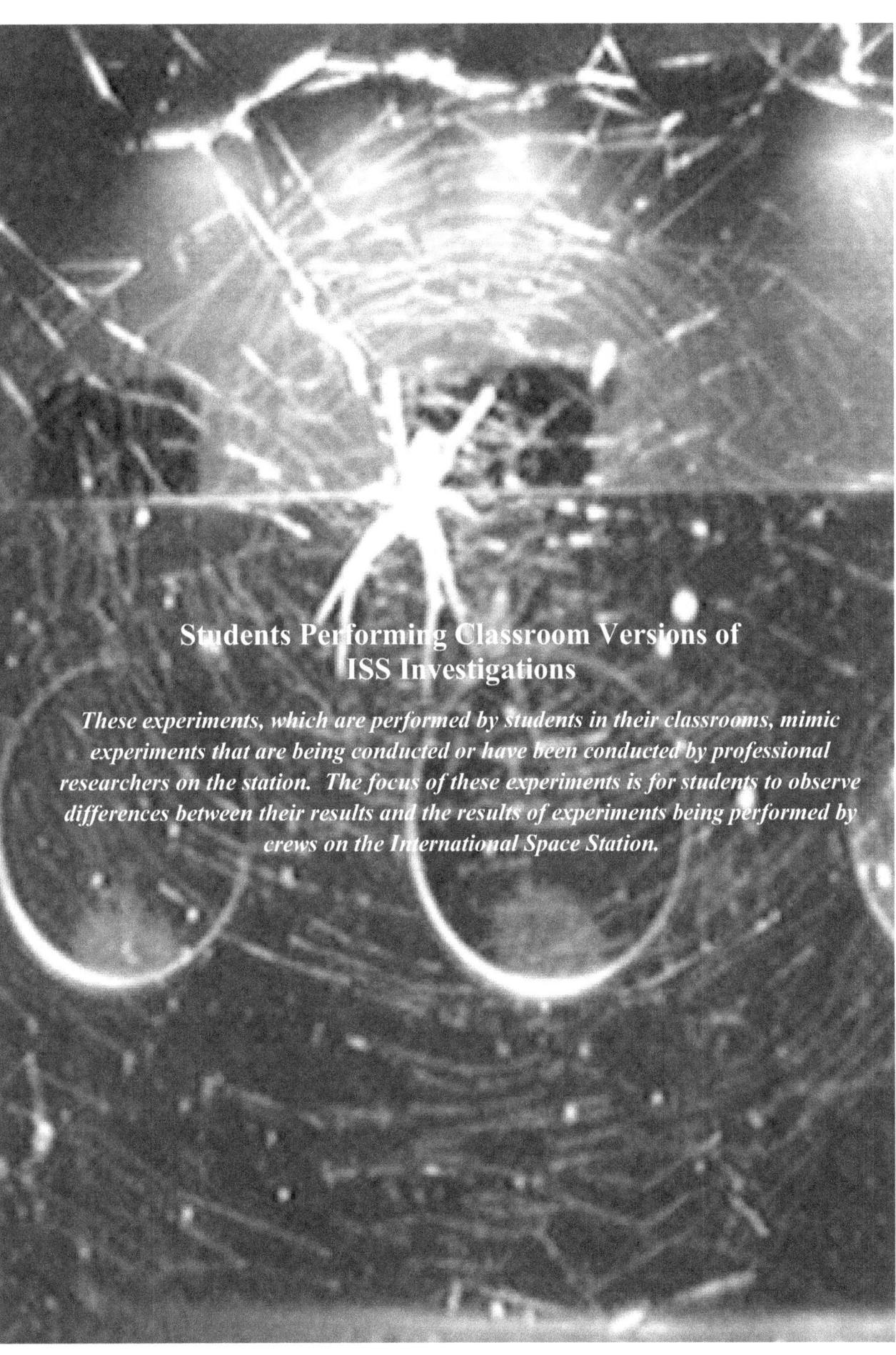

Students Performing Classroom Versions of ISS Investigations

These experiments, which are performed by students in their classrooms, mimic experiments that are being conducted or have been conducted by professional researchers on the station. The focus of these experiments is for students to observe differences between their results and the results of experiments being performed by crews on the International Space Station.

Commercial Generic Bioprocessing Apparatus Science Insert – 05: Spiders, Fruit Flies and Directional Plant Growth (CSI-05)

Expeditions: 27/28, 29/30

Leading Space Agencies: NASA, Global

Curriculum Grade Levels: K–8 (elementary), 9–12 (secondary), Undergraduate (college, postsecondary), Graduate (master's, Ph.D., M.D.)

Impact: More than 119,000 students and 2000 teachers and schools worldwide have participated in this classroom version of the International Space Station experiments.

Participating Countries: Australia, Brazil, Canada, Chile, China, Colombia, Ecuador, Egypt, Germany, Hong Kong, India, Iran, Ireland, Italy, Japan, Mexico, Pakistan, Panama, Poland, Romania, Russia, South Africa, Switzerland, Taiwan, United Arab Emirates, United Kingdom, United States

Participating States: Alabama, Alaska, Arizona, Arkansas, California, Colorado, Connecticut, District of Columbia, Delaware, Georgia, Florida, Hawaii, Idaho, Illinois, Indiana, Iowa, Kansas, Louisiana, Maine, Maryland, Massachusetts, Michigan, Minnesota, Mississippi, Missouri, Montana, North Carolina, Nebraska, Nevada, New Hampshire, New Jersey, New Mexico, New York, Ohio, Oklahoma, Oregon, Pennsylvania, Puerto Rico, Rhode Island, South Carolina, South Dakota, Tennessee, Texas, Utah, Vermont, Virginia, Washington, West Virginia, Wyoming

Number of K–8 Students (elementary): 106,416

Number of 9–12 Students (secondary): 13,009

Number of Undergraduate Students (college, postsecondary): 8

Number of Graduate Students (master's, Ph.D., M.D.): 8

Number of Schools: 1989

Number of Teachers: >2000

Description of Student Participation and Activities: The Commercial Generic Bioprocessing Apparatus Science Insert (CSI) program provides the K–12 community opportunities to use the unique microgravity environment of the International Space Station as part of the regular classroom to encourage learning and interest in STEM. CSI-05 examines the web spinning characteristics of an orb weaving spider and the motility behavior of the fruit fly.

One of the golden orb spiders aboard the International Space Station. Image courtesy of NASA.

The students follow curriculum designed for the classroom that examines how the organisms eat, grow, molt, spin webs and move in space. Many classrooms are provided a classroom kit with the organisms so that students are able to compare what is happening on the ISS experiment with the experiment in their classroom. Classrooms that did not receive an official classroom kit are provided instructions on how to build their own kit and secure their own organisms. Daily images and video are downlinked from the station and uplinked to the bioedonline.org website, providing students easy access to the experiment data. The second half of CSI-05 is a plant experiment, which began in September 2011.

Description of Teacher Participation and Activities: Teachers facilitated all classroom activities.

Education Lead: Stefanie Countryman, BioServe Space Technologies, Boulder, Colorado, USA

Education Websites: http://bioedonline.org/, http://www.colorado.edu/engineering/BioServe/

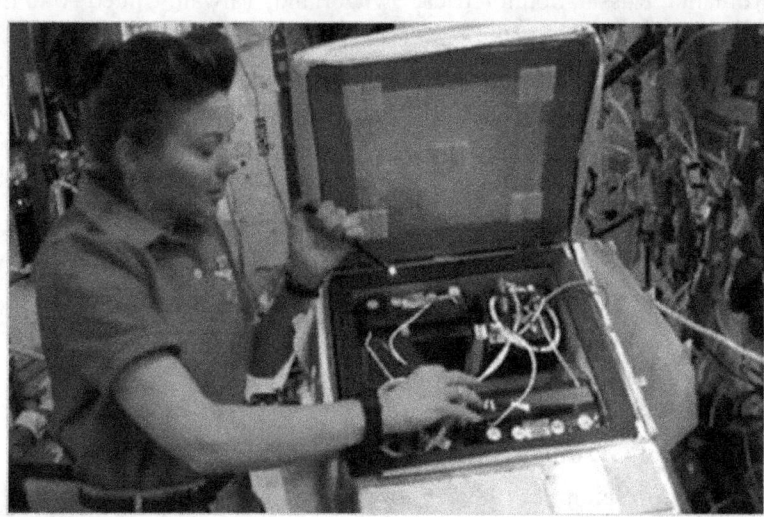

NASA astronaut Catherine (Cady) Coleman shows the spider habitat. Image courtesy of NASA.

Students Participating in ISS Investigator Experiments

Many research investigators have enlisted students to help with their experiments. Some of these experiments are performed solely to inspire the next generation of scientists, engineers and explorers. Others involve undergraduate and graduate students as well as postdoctoral fellows who are working on research under the guidance of investigators and university professors.

Crew Earth Observations (CEO)

Expeditions: 1–33/34, Ongoing
Leading Space Agencies: NASA, Global
Curriculum Grade Levels: 3–12, College (undergraduate)

Participating Country: United States

Number of K–8 Students: 11,000
Number of 9–12 Students: 10,500
Number of Undergraduate Students (college, postsecondary): 1000
Number of Graduate Students (master's, Ph.D., M.D.): 1000
Number of Teachers: 5000

Experiment Description: The goal of CEO is to obtain qualitative and quantitative digital photographs of Earth for use in educational and scientific applications. The imagery is taken by space station crews using digital cameras over specified regions of the Earth. CEO site-selection criteria include indicators of global change, conservation management monitoring sites, terrestrial analogs of features found on other planetary surfaces, human settlements and events demonstrating the dynamism of our planet. Examples of selected targets include river deltas, glaciers and water level changes in lakes (global change); ecological preserves and coral reefs (conservation);

http://eol.jsc.nasa.gov/EarthObservatory/OmahaCouncilBluffsLewisClarkTrail.htm

The labeled image at right is a subset excerpt of the image above.

In 2003, these images were posted on the Goddard Space Flight Center Earth Observatory website on a Lewis and Clark Geosystem Portal, a NASA Stennis-sponsored teacher support site run by GCS Research, LLC, at the University of Montana – Missoula.

impact craters and volcanic fields (planetary studies); cities, agriculture and explorer trail geography (human settlements); and large storms, forest fires or dust storms (dynamic events). Image data resulting from the CEO experiment are distributed via the publicly accessible Gateway to Astronaut Photography found at http://eol.jsc.nasa.gov, which contains more than 185,000 images of the Earth taken from the International Space Station. The database, which is accessed by educators, students, scientists and the general public from over 130 countries, supports roughly 750,000 image downloads per month.

Description of Student Participation and Activities: Over the life of the International Space Station Program, the CEO staff has conducted or participated in numerous educational venues, such as distance learning lectures, in-class or school projects or lectures, onboard educational video material, career day conferences, and student intern programs. In more focused efforts, such as Paige Graff's project, Expedition Earth and Beyond, the CEO staff has engaged the students and taught the process of science using crewmember Earth Observations photography. The direct activities include conducting hands-on image interpretation, taking measurements from imagery, relating maps to imagery, and learning many aspects of geography. In 2003, students used the interactive displays in museums across the country, where they worked together and provided imagery to celebrate the 200th year of the Lewis and Clark expedition.

Description of Teacher Participation and Activities: Over the life of the International Space Station Program, the CEO staff has conducted or participated in numerous educational venues, such as teacher workshops, distance learning lectures, in-class or school projects or lectures, onboard educational video material, career day conferences, and student intern programs; attended educational conferences; and in more focused efforts, such as Expedition Earth and Beyond, engaged students and taught the process of science using astronaut Earth Observations photography.

More recently, Expedition Earth and Beyond participants at Charleston Middle School successfully completed studies of coral reefs. They selected this image (ISS027-e-14219) as a reward for completing the research objectives.

Mozambique Channel, Juan de Nova Island and Reef (17.0S 42.7E).

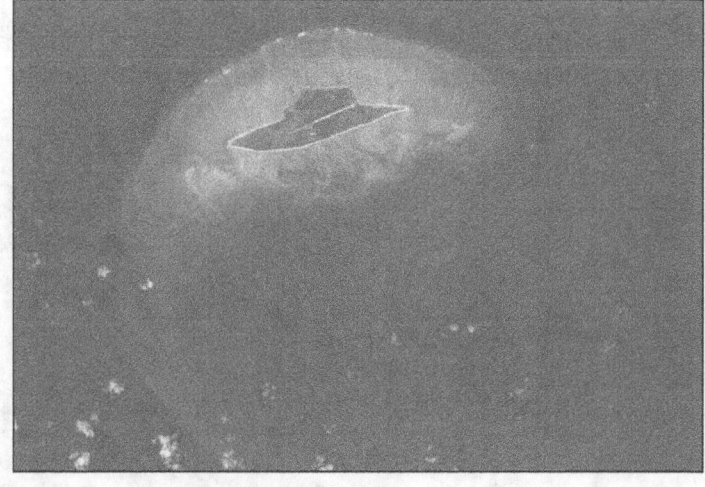

Principal Investigators: Kim Willis and Paige Graff, NASA JSC, Houston, Texas, USA
Education Websites:
Gateway to Astronaut Photography of Earth (GAPE) – http://eol.jsc.nasa.gov
Expedition Earth and Beyond – http://ares.jsc.nasa.gov/ares/eeab/index.cfm

Student Involvement in Other ISS Investigations

As information was being gathered for this document, International Space Station principal investigators were asked to supply information about student involvement in their research. A sustained and cumulative impact of ISS research was reported by many investigators. There is likely additional student involvement in projects that did not respond to our survey. In most cases, the responding investigators had invited university students at a variety of levels to participate in some aspect of their ISS research. Investigations with associated student activities that have not been included elsewhere in this document are summarized below. First the basic objectives of the research and then the student participation are described. Numbers of participating students are summarized in Table 10.

Table 11 – Student Involvement in Other International Space Station Investigations

Investigations	Number of Students				Schools	Teachers
	K-12	Undergraduate	Graduate	Postdoctoral		
Analysis of a Novel Sensory Mechanism in Root Phototropism (Tropi)		5	3	2		
Binary Colloidal Alloy Test-3 Critical Point (BCAT-3-CP)		1	1	1		
Capillary Channel Flow (CCF)		1	3	1	1	1
Capillary Flow Experiments (CFE)	10,100	200	14	1	2	3
Cardiovascular and Cerebrovascular Control on Return from International Space Station (CCISS); Cardiovascular Health Consequences of Long-Duration Space Flight (Vascular)		100	12	3	2	3
Constrained Vapor Bubble (CVB)		3	7		2	2
Coulomb Crystal						
Device for the Study of Critical Liquids and Crystallization – Directional Solidification Insert (DECLIC-DSI)			3	4		
ELaboratore Immagini TElevisive – Space 2 (ELITE-S2)		1	1	1		

Investigations	Number of Students				Schools	Teachers
	K-12	Undergraduate	Graduate	Postdoctoral		
Hyperspectral Imager for Coastal Ocean (HICO) and Remote Atmospheric and Ionospheric Detection System (RAIDS) Experiment Payload – Remote Atmospheric and Ionospheric Detection System (HREP-RAIDS)		5	2	1	6	3
Integrated Resistance and Aerobic Training Study (SPRINT)			2	1	2	
Investigating the Structure of Paramagnetic Aggregates from Colloidal Emulsions–2 (InSPACE-2)	1	1	1	3		
Materials International Space Station Experiment – 7 (MISSE-7)		5				1
Mental Representation of Spatial Cues During Spaceflight (3D-Space)			5			
Microheater Array Boiling Experiment (MABE): Flight Research Using the Boiling Experiment Facility (BXF)		9	3	2		
Nucleate Pool Boiling Experiment (NPBX)	6	6	8	4	1	2
Nutritional Status Assessment (Nutrition) and Dietary Intake Can Predict and Protect Against Changes in Bone Metabolism during Spaceflight and Recovery (Pro K)		7	20	1		
Physiological Factors Contributing to Changes in Postflight Functional Performance (Functional Task Test)		5	6	1	8	1
Psychomotor Vigilance Self-Test on the ISS (Reaction Self Test)		3			2	
Sensor Test for Orion Relative Navigation Risk Mitigation (STORRM)		12	1		8	

Investigations	Number of Students				Schools	Teachers
	K-12	Undergraduate	Graduate	Postdoctoral		
Space Communication and Navigation Testbed (SCAN TESTBED)						
Space-Dynamically Responding Ultrasound Matrix System (SPACE DRUMS)			7		1	1
Total	10,107	364	99	26	35	17

Analysis of a Novel Sensory Mechanism in Root Phototropism (Tropi)
John Z. Kiss, Ph.D., Miami University, Oxford, Ohio, USA

Tropi is a plant growth experiment that investigates how plant roots from *Arabidopsis thaliana* (thale cress) respond to varying levels of light and gravity. The plants grown are analyzed to determine which genes are responsible for successful plant growth in microgravity. This experiment provides insight into the growth of plants in space, which will help scientists and engineers create sustainable life-support systems for long-term space travel.

Students were involved in ground-based baseline studies, mission operations during station experiments and postflight data analyses.

NASA astronaut Jeffrey Williams, Expedition 22 Commander, services the Tropi experiment in the Columbus Laboratory of the International Space Station. NASA image ISS022E087465

Binary Colloidal Alloy Test-3 and -4: Critical Point (BCAT-3-4-CP)
David Weitz, Ph.D., and Peter Lu, Ph.D., Harvard University, Cambridge, Massachusetts, USA

In the BCAT-3-4-CP experiment, crewmembers photograph samples of polymer and colloidal particles (tiny nanoscale spheres suspended in liquid) that model liquid-gas phase changes. The results help scientists understand the phase behavior of a model colloid-polymer system that is near its critical point, the point in atomic systems at which the material has both liquid and gas properties.

Several graduate students have assisted with training crewmembers and designing the science and have prepared the samples and presented the results.

Expedition 12 Commander and science officer William (Bill) McArthur photographs BCAT-3 experiment samples. NASA image ISS012E07685

Capillary Channel Flow (CCF)
Michael Dreyer, Ph.D., University of Bremen, Bremen, Germany

The CCF experiment helps scientists design innovate ways to transport liquids (e.g., fuels and low-temperature liquids, such as liquid nitrogen and water) in microgravity. By understanding capillary fluid flow in microgravity, hardware can be developed to pump liquids from one reservoir to another without the need for a pump with moving parts. The reduced cost and weight and improved reliability of such equipment make this a particularly attractive technology for NASA.

Students perform the experiment (operations on the station), evaluate experiments, and perform computations with 3D and 1D models. The teachers lead the investigation, coordinate the different activities, and maintain contact with agencies (e.g., NASA and DLR) and with the payload developer.

Capillary Flow Experiments (CFE)

Mark Weislogel, Ph.D., Portland State University, Portland, Oregon, USA

Capillary Flow Experiments (CFE) is a suite of fluid physics experiments that investigates how fluids move up surfaces in microgravity. The results are meant to improve current computer models that are used by designers of low-gravity fluid systems and may improve fluid transfer systems for water on future spacecraft. Wetting describes the ability of a liquid to spread across a surface. Understanding how wetting occurs in microgravity is important for a wide variety of engineering systems being developed for human space flight. CFE studies may demonstrate how liquids can be separated from gas in the absence of gravity — a key procedure for water purification systems in microgravity.

Students attend seminars, give class presentations, launch high-altitude balloons and perform aircraft piggyback experiments. They also design and conduct drop tower experiments and have summer research opportunities. In addition to the activities listed above, undergraduate students design projects related to drop towers, aircraft and space experiments. They assist with experiment design as well as data reduction and presentation. Graduate students are involved in all of the above activities as well as crew procedures development, crew training, flight experiment control, data collection and reduction, presentation, thesis completion and publication. Postdoctoral students are involved in all of the above, plus they mentor graduate students and assist in course preparation and presentation.

NASA astronaut Catherine (Cady) Coleman, Expedition 26 flight engineer, performs a CFE Interior Corner Flow 2 (ICF-2) test. NASA image ISS026E018760

Cardiovascular and Cerebrovascular Control on Return From International Space Station (CCISS); Cardiovascular Health Consequences of Long-Duration Space Flight (Vascular)

Richard L. Hughson, Ph.D., University of Waterloo, Waterloo, Canada

The CCISS investigation studies the effects of long-duration space flight on crewmembers' heart functions and the blood vessels that supply the brain. Learning more about the cardiovascular and cerebrovascular systems could lead to specific countermeasures that might better protect future space travelers. This experiment is a collaborative effort with the CSA.

Approximately 30 to 40 undergraduate and graduate students have an opportunity each year to take a course in microgravity hypo- and hyperbaric physiology. The major part of the course focuses on microgravity physiology and CCISS and Vascular data collection experiences. Students are also provided an opportunity to see a demonstration laboratory that contains equipment used to collect data on the crewmembers. Four students have been extensively involved in CCISS data analysis. One student participated in data collection at Dryden Space Flight Center, where she processed the Holter and Actiwatch data from all crewmembers. She has also presented data at several conferences and was the first author on a manuscript recently submitted to the Journal of Applied Physiology.

Constrained Vapor Bubble (CVB)

Peter C. Wagner, Ph.D., and Joel L. Plawsky, Sc.D., Rensselaer Polytechnic Institute, Troy, New York, USA

Constrained Vapor Bubble (CVB) aims to achieve a better understanding of the physics of evaporation and condensation and how they affect cooling processes in microgravity using a remotely controlled microscope and a small cooling device. Certain types of cooling devices, known as wickless heat pipes, contain no moving parts and, as a result, are highly reliable in spaces where access to replacement parts is difficult or impossible. The results from these experiments could lead to the development of more efficient cooling systems in microelectronics on Earth and in space.

Students were involved in all phases of the experimental and theoretical program. They provided the ground-based testing to verify the experimental configuration and developed the theory and models needed to analyze the data. Two spent a considerable amount of time at Glenn Research Center helping to build and test the apparatus. In addition, one student was primarily responsible for being at Glenn Research Center during ISS operations, ensuring that needed data was obtained. This research and investigation involved the participation of several doctoral students. Seven doctoral students associated with this research received their Ph.D. degrees.

Coulomb Crystal

V. E. Fortov, Academician, Joint Institute for High Temperatures of the Russian Academy of Sciences (JIHT RAS), Russia

The Coulomb Crystal investigation gives graduate and postgraduate students the opportunity to prepare and conduct the experiment to study the dynamics of solid and dispersed environments in inhomogeneous magnetic fields in microgravity.

The Coulomb Crystal experiment includes sessions in the microgravity properties of dispersed dust, Coulomb crystals and Coulomb's liquids, which form charged macro particles in a magnetic trap. Pilot studies on board the ISS explore the structural properties of Coulomb clusters, liquid crystal phase transitions. They also examine the wave processes; thermal, physical and mechanical characteristics; and transport phenomena in Coulomb systems of charged particles. These charged particles are found in a magnetic field.

Website: http://knts.tsniimash.ru/en

Coulomb Crystal equipment: electromagnetic set with plug-in container, including model material. Image courtesy of Roscosmos.

Buildup of the ellipsoidal form Coulomb cluster in the magnetic trap. The particle size is 300 micrometers. Image courtesy of Roscosmos.

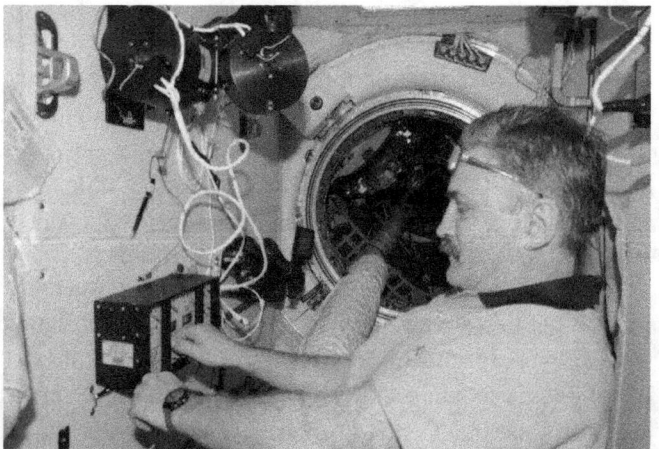

The Coulomb Crystal scientific experiment is implemented by the Russian cosmonaut A. Yu. Kaleri (onboard engineer on Expedition 25/Expedition 26). Image courtesy of Roscosmos.

Device for the Study of Critical Liquids and Crystallization – Directional Solidification Insert (DECLIC-DSI)

Nathalie Bergeon, Ph.D., Université Paul Cézanne (Aix-Marseille III), Marseille, France; Rohit Trivedi, Ph.D., Ames Laboratory, U.S. Department of Energy, Iowa State University, Ames, Iowa, USA

DECLIC-DSI involves the in situ and real-time observation of the microstructures that form at the liquid-solid interface when transparent materials solidify. This makes it possible to simulate metallic alloys with the advantage of recording the dynamic formation and selection of the microstructures. Another practical advantage is that images, instead of heavy samples, are transferred to the ground for post-mortem characterization, since the transport into orbit and the return to Earth greatly increases the cost of missions.

Students are involved in designing a laboratory model and testing the concepts. For ground experiments, students are involved in choosing and testing different alloys, analyzing solidification experiments on the laboratory device in a gravity environment, and improving the instrument. For flight experiments, students are involved in defining and realizing the flight experiments and analyzing the results.

ELaboratore Immagini TElevisive – Space 2 (ELITE-S2)
Giancarlo Ferrigno, Ph.D., Politecnico di Milano, Milano, Italy

ELITE-S2 investigates the connection between the brain, visualization and motion in the absence of gravity. By recording and analyzing the 3D motion of crewmembers, this study helps engineers apply ergonomics to future spacecraft designs and determines the effects of weightlessness on breathing mechanisms for long-duration missions. The experiment was designed and performed in cooperation with ASI.

On Earth, the ability to catch a ball depends on a mental model of the physical behavior of that object, a model that includes gravity. In a microgravity environment, crewmembers adjust their motor control strategies to respond to new rules, but still show evidence that the old gravity-based rules are hard wired into their brains through neural networks. This experiment evaluates differences in the way the brain controls conscious and unconscious motions, such as breathing, sitting and standing in environments with and without gravity.

A student with a master's degree in bioengineering from Politecnico di Milano, NearLab, developed his final project for the first campaign of ELITE-S2 on board the International Space Station; his activity was focused on kinematic and dynamic analyses of data coming from two crewmembers involved in Movement in Orbital Vehicle Experiments (MOVE) acquisitions (whole body pointing movements). A doctoral student followed and supported the pre-flight, in-flight and postflight sessions of ELITE-S2; she joined the training and the Base Data Collection (BDC) pre-flight and postflight sessions at JSC in Houston, Texas. She supported the in-flight experiments through the control base station and was involved primarily in the data analysis and interpretation. A postdoctoral student was involved mainly in the preliminary phases of protocol and procedure definition and the setup testing.

Hyperspectral Imager for Coastal Ocean (HICO) and Remote Atmospheric and Ionospheric Detection System (RAIDS) Experiment Payload – Remote Atmospheric and Ionospheric Detection System (HREP-RAIDS)
Scott Budzien, Naval Research Laboratory, Washington, D.C., USA

The HREP-RAIDS experiment combines two experiment sensors into one payload and provides atmospheric scientists with a complete description of the major constituents of the thermosphere and ionosphere. The thermosphere is the layer of the Earth's atmosphere where the International Space Station orbits the Earth, and the ionosphere is the portion of the upper atmosphere that affects radio waves. RAIDS provides density, composition, temperature and electron density profiles at altitudes between 95 and 300 kilometers.

A selection of HICO best images.

Undergraduate students' participation has ranged from assembling and testing flight hardware to developing ground system software and performing science data analysis. The RAIDS data set has been distributed to a number of co-investigators who have involved undergraduate, graduate, and postdoctoral students in science analysis and conference presentations.

Integrated Resistance and Aerobic Training Study (SPRINT)

Lori Ploutz-Snyder, Ph.D., NASA JSC, Houston, Texas, USA

SPRINT evaluates the use of high-intensity, low-volume exercise training to minimize loss of muscle, bone, and cardiovascular function in station crewmembers during long-duration missions. Current exercise countermeasures on the International Space Station are insufficient to prevent muscle atrophy, cardiovascular deconditioning and bone loss associated with long-duration space flight. There is a need to prevent space-flight-related deconditioning to protect the health and mission readiness of the current crew as well as to enable NASA to protect the fitness of crewmembers who anticipate missions of even longer duration to the moon and Mars. The investigator's long-range goal in the Exercise Physiology and Countermeasures Project is to develop and optimize exercise countermeasures for use in long-duration space flights.

For the elementary grades (K–8), the team works with NASA education as a content area expert in development of Train Like an Astronaut, which uses examples and information from the SPRINT study. At the secondary level (grades 9–12), presentations about the SPRINT study are given to high school audiences in support of careers in math and science. Doctoral students and postdoctoral fellows are involved in the actual SPRINT data collection.

Investigating the Structure of Paramagnetic Aggregates from Colloidal Emulsions–2 (InSPACE-2)
Eric M. Furst, Ph.D., University of Delaware, Newark, Delaware, USA

InSPACE-2 studies the fundamental behavior of magnetic colloidal fluids under the influence of various magnetic fields. Observations of the microscopic structures yield a better understanding of the interplay of magnetic, surface and repulsion forces between structures in magnetorheological fluids. These fluids are classified as smart materials that transition to a solid-like state by forming and cross-linking microstructures in the presence of a magnetic field. On Earth, these materials are used for vibration damping systems that can be turned on or off.

Secondary and postsecondary students, working with the principal investigator and postdoctoral researchers, designed and executed ground-based experiments and aided in processing ISS experimental data.

Materials International Space Station Experiment – 7 (MISSE-7)
James R. Grier, NASA Glenn Research Center, Cleveland, Ohio, USA

MISSE-7 is a suite of experiments that includes more than 700 new and affordable materials. The samples tested have the potential for use in advanced reusable launch systems and advanced spacecraft systems, including solar cells, optics, sensors, electronics, power, coatings, structural materials and protection for the next generation of spacecraft. The development of the next generation of materials and material technologies is essential to the mission of traveling beyond Earth's orbit. Undergraduate students helped with pre-flight and postflight analysis of the MISSE-7 Spacesuit Fabric Exposure Experiment.

Mental Representation of Spatial Cues During Space Flight (3D-Space)
Prof. Gilles Clement, Centre National de la Recherche Scientifique, Toulouse, France

The Mental Representation of Spatial Cues During Space Flight (3D-Space) experiment investigates the effects of exposure to microgravity on the mental representation of spatial cues by crewmembers during and after space flight. The absence of the gravitational frame of reference during space flight could be responsible for disturbances in the mental representation of spatial cues, such as the perception of horizontal and vertical lines, an object's depth, and a target's distance.

Senior scientists participated with the students in the Mice Drawer System (MDS) experiment, leading the research and providing all required advice to their younger colleagues.

Microheater Array Boiling Experiment (MABE):
Flight Research Using the Boiling Experiment Facility (BXF)
Jungho Kim, Ph.D., University of Maryland, College Park, Maryland, USA

Boiling efficiently removes large amounts of heat by generating vapor from liquid; this process is currently being used in many power plants to generate electricity. An upper limit, called the critical heat flux, exists when the heater is covered with so much vapor that liquid supply to the heater begins to decrease, potentially destroying the heater. MABE determines the critical heat flux during boiling in microgravity to design optimal cooling systems for future space exploration vehicles as well as for use on Earth.

The students design, build and operate experiments for the low-gravity aircraft, then analyze the data. They must perform stress analysis to meet aircraft requirements, design and build control hardware and data acquisitions systems, acquire and analyze the data, compare the data with data from existing models and develop new models as necessary.

Nucleate Pool Boiling Experiment (NPBX)
Vijay Dhir, Ph.D., University of California Los Angeles, Los Angeles, California, USA

Nucleate boiling is bubble growth from a heated surface and the subsequent detachment of the bubble to a cooler surrounding liquid (bubbles in microgravity grow to different sizes than bubbles on Earth). As a result, these bubbles can transfer energy through fluid flow. The NPBX investigation provides an understanding of heat transfer and vapor removal processes that take place during nucleate boiling in microgravity. This understanding is needed for optimum design and safe operation of heat exchange equipment that uses nucleate boiling as a way to transfer heat in extreme environments of the deep ocean (submarines) and microgravity.

Students participated in ground-based experiments of bubble dynamics and nucleate boiling heat transfer. They also developed numerical simulation of the process and used the simulation tool to predict the influence of reduced gravity, including the level of gravity on the ISS.

This is the first time simulation of the boiling process in microgravity has been carried out prior to experiments.

Nutritional Status Assessment (Nutrition) and Dietary Intake Can Predict and Protect Against Changes in Bone Metabolism during Spaceflight and Recovery (Pro K)

Scott M. Smith, Ph.D., NASA JSC, Houston, Texas, USA

Nutritional Status Assessment (Nutrition) is a comprehensive in-flight study designed to understand changes in human physiology during long-duration space flight. This study includes measures of bone metabolism, oxidative damage, and chemistry and hormonal changes as well as assessments of the nutritional status of the crewmembers participating in the study. The results affect the definition of nutritional requirements and development of food systems for future exploration missions to the moon and Mars. This experiment also helps researchers understand the effectiveness of measures taken to counteract the effects of space flight and the impact of exercise and pharmaceutical countermeasures on nutritional status and nutrient requirements for crewmembers.

Approximately 27 undergraduate and graduate students have worked in the nutrition laboratory during the past five years [since 2006, which matches the start of our space station experiments Nutrition Supplemental Medical Objective (SMO) and Stability, with Pro K beginning in 2008]. These students have been involved in four-week to three-month internships with the laboratory. One of the students is finishing a Master of Science thesis at Texas Woman's University, analyzing bed-rest data (with a distinct Pro K spin). Countless tours; briefings; seminars; and local, national, and international presentations have been supported by this work. A Space Nutrition Newsletter was developed in 2001 as a way to engage children in the upper elementary and intermediate grades. This continued through approximately 2007. All materials are available online at http://www.nasa.gov/centers/johnson/slsd/about/divisions/hacd/education/kids-zone.html.

In 2011, work was done to expand content from the Space Nutrition Newsletters to create a book, which is currently in manuscript form.

Food cans and packets float freely on board ISS during Expedition 7. A balanced meal is important to the overall nutrition and health of the crew during long-duration exploration. NASA image ISS007E06700

Physiological Factors Contributing to Changes in Postflight Functional Performance (Functional Task Test)

Jacob Bloomberg, Ph.D., NASA JSC, Houston, Texas, USA

The Physiological Factors Contributing to Changes in Postflight Functional Performance (Functional Task Test) payload tests crewmembers on an integrated suite of functional and physiological tests before and after short- and long-duration space flight. The study identifies critical mission tasks that may be impacted, maps physiological changes to alterations in physical performance and aids in the design of countermeasures that specifically target the physiological systems responsible for impaired functional performance. Students participated in supporting ground-based studies for this investigation.

Psychomotor Vigilance Self Test on the International Space Station (Reaction Self Test)

David F. Dinges, Ph.D., University of Pennsylvania School of Medicine, Philadelphia, Pennsylvania, USA

Reaction Self Test is a portable, five-minute reaction time task that allows crewmembers to monitor the daily effects of fatigue on performance while on board the International Space Station. It aids crewmembers in objectively identifying when their performance capability is degraded by various fatigue-related conditions that can occur as a result of ISS operations and time in space (e.g., acute and chronic sleep restriction, slam shifts, extravehicular activity (EVA), and residual sedation from sleep medications). The Reaction Self Test also evaluates the extent to which the performance of station crewmembers is sensitive to fatigue from sleep loss and circadian [a rhythm of biological functions occurring in a 24-hour periodic cycle (e.g., sleeping and eating)] disruption during the mission, fatigue from work intensity during the mission, decline of performance with time during the mission, and carryover effects of medications for sleep on board the ISS. The Reaction Self Test also evaluates the extent to which performance feedback (via a graphical interface) is perceived by crewmembers as a useful tool for assessing performance capability.

Undergraduate students worked to develop a computer calibrator for crewmember computers used in Reaction Self Test data acquisition as well as to develop a timeline of space station activities related to Reaction Self Test data acquisition for study. In addition, they worked to summarize and plot Reaction Self Test data from the ISS.

Sensor Test for Orion Relative Navigation Risk Mitigation (STORRM)
Heather Hinkel, NASA JSC, Houston, Texas, USA

STORRM tested the Vision Navigation Sensor, Star Tracker, and docking camera planned for Orion during the space shuttle's approach to and departure from the International Space Station. This test determined how well the navigation system performed during the mission.

The STORRM project flew a cutting-edge relative navigation sensor on the shuttle and tested it on a mission to ISS. Students assisted with developing tools for data analysis, performed data analysis, performed alignment calculations, and assisted with mission support activities.

Space Communications and Navigation Testbed (SCAN Testbed)
Richard Reinhart, NASA Glenn Research Center, Cleveland, Ohio, USA

The SCAN Testbed consists of reconfigurable software-defined radios with software-based communications and navigation functions that provide ground mission planners the ability to change the functionality of the radio on-orbit. The ability to change the operating characteristics of the radio's software after launch allows missions to change the way a radio communicates with ground controllers and offers flexibility to adapt to new science opportunities and increased data return. Also, this flexibility allows teams to recover from anomalies in the science payload or communication system and potentially reduce development cost and risk by using the same hardware platform for different missions while using software to meet specific mission requirements.

Students at any level conceive an experiment with the flight system. For graduate and postgraduate students, experiments include developing and providing software to run on the software defined radios for the flight system and the ground system. Investigators are expected to secure development funding and publish their results. For younger students, an experiment that explains orbits and communications and perhaps sends pictures to space and back is conceivable.

Space Dynamically Responding Ultrasonic Matrix System (SpaceDRUMS)

Jacques Guigne, Ph.D., Guigne Space Systems, Inc., Paradise, Newfoundland, Canada

The goal of Space Dynamically Responding Ultrasonic Matrix System (SpaceDRUMS) is to provide a suite of hardware capable of facilitating containerless advanced materials science, including combustion synthesis and fluid physics. Inside SpaceDRUMS®, samples of experimental materials can be processed without touching a container wall.

Graduate students worked on research associated with developing the science for this investigation.

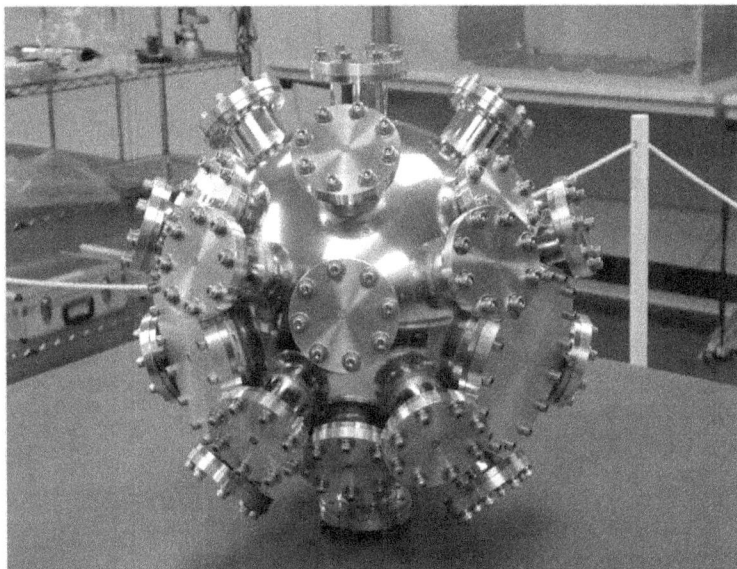

SpaceDRUMS processing chamber. Image courtesy of CSA.

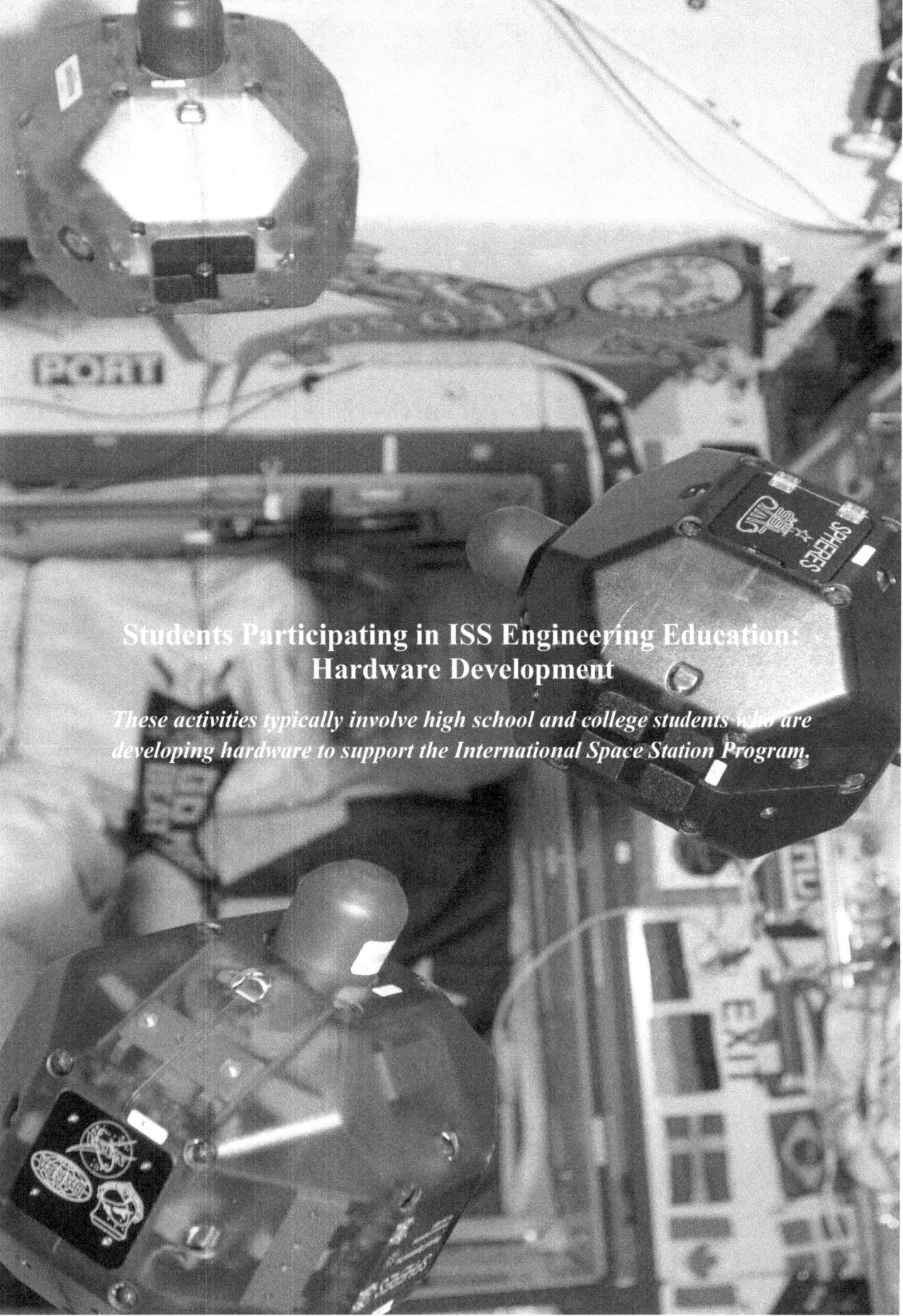

Students Participating in ISS Engineering Education: Hardware Development

These activities typically involve high school and college students who are developing hardware to support the International Space Station Program.

High School Students United With NASA to Create Hardware (HUNCH)

Expeditions: 2003–present

Leading Space Agency: NASA

Curriculum Grade Levels: 7–8 (elementary), 9–12 (secondary)

Impact: Almost 1500 students from 12 states have been involved in HUNCH projects.

Participating Country: United States

Participating States: Alabama, Colorado, Kentucky, Louisiana, Massachusetts, Mississippi, Montana, Pennsylvania, South Dakota, Tennessee, Texas, Wyoming

Number of Students: 1488

Number of Teachers: 51

Description of Student Participation and Activity: HUNCH is an innovative, school-based program that joins NASA with high schools and middle schools in states across the nation. The partnership involves students fabricating real-world products for NASA as they apply their science, technology, engineering and mathematics skills and learn to work in teams and think creatively. Career and technical education courses provide the perfect settings for the HUNCH program.

NASA receives cost-effective hardware, soft goods and educational videos that are produced by the students. The students receive hands-on experience and, in some cases, NASA certification in the development of training hardware for International Space Station crewmembers or ground support personnel. NASA provides the materials required for building the hardware and soft goods along with drawings and other documents needed to fabricate the items. NASA also provides quality inspection oversight during the fabrication of the hardware. The school provides technical direction, provides a safe working environment, teaches the students how to use the tools and provides pictures of the process.

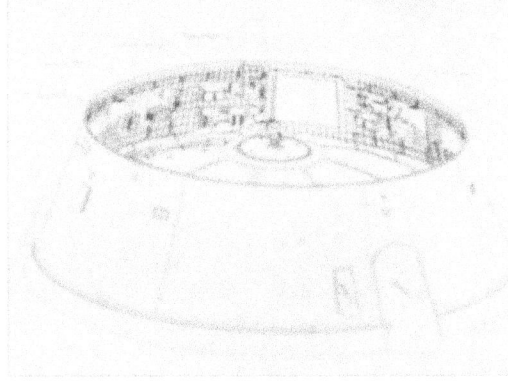

Lusher students Kendall LaSalle, left, and David Lopez with the drawings and finished model of an Ortho Grid, which the class team created. The Ortho Grid is a storage area on the International Space Station. Image courtesy of Eliot Kamenitz, The Times-Picayune.

Education Lead: Stacy L. Hale, NASA JSC, Houston, Texas, USA

Education Website: www.nasahunch.com

"I was already interested in engineering, and this confirmed I was good at it," said Kendall LaSalle, a student at Lusher Charter School who hopes to attend Baylor University in Texas.

With knowledge gained in a concurrent geometry class, LaSalle was able to apply theorems to solve the real life, spatial problems posed by the NASA project. "I learned the basics of engineering — what you have to observe about what you're going to make," he said.

International Space Station Agricultural Camera (ISAAC)

Expeditions: 27/28–33/34, Ongoing

Leading Space Agency: NASA

Curriculum Grade Levels: Undergraduate (college, postsecondary), Graduate (master's, Ph.D., M.D.)

Participating Country: United States

Participating State: North Dakota

Number of Undergraduate Students (college, postsecondary): 41

Number of Graduate Students (master's, Ph.D., M.D.): 23

Experiment Description: ISSAC is a multispectral camera used on the ISS as a subrack payload of the Window Observational Research Facility (WORF). It takes frequent images in visible and infrared light, principally of vegetated areas (growing crops, grasslands, forests) in the northern Great Plains region of the United States. The sensor also is being used to study dynamic Earth processes around the world, such as melting glaciers, ecosystem responses to seasonal changes, human impacts, and rapid-response monitoring of natural disasters.

Description of Student Participation and Activities: ISSAC was designed, developed, tested, and is being operated by university graduate and undergraduate students at the University of North Dakota. To date, 64 students from eight different departments across campus have contributed to the project. The students' technical contributions have been made predominately through graduate research assistantships, although there have been only 13 of these to date. Of these, nine were thesis and four were non-thesis graduate projects. Almost half of the students have been compensated in an hourly position; most of these have been undergraduates, although a few were graduate students at the time of their employment. Contributions from a number of students came through their participation in a Research Experiences for Undergraduates (REU) project funded by the National Science Foundation. A few students, such as those who participated as part of an undergraduate senior design project or graduate-level independent research topic, participated for academic credit only.

Education Lead: Doug Olsen, University of North Dakota, Grand Forks, North Dakota, USA

Education Website: www.umac.org/issac

> *George Seielstad, Director of AgCam and the University of North Dakota Center for People and Environment: "AgCam provides students with the opportunity to do real engineering and provide valuable data to protect our environment now and in the future."*

Students at the University of North Dakota conducting on-orbit operations at the ISSAC Science Operations Center on the campus of the University of North Dakota. Front (left to right) Miyuru Arangala, a computer science student, and Scott Arbuckle, a commercial aviation and entrepreneurship student. Back (left to right) David Butz, a commercial aviation student, and Bhanu Chennamaneni, a computer science student.

AgCam components installed in the Window Observational Research Facility (WORF) ground test rack are shown with the installation of EarthKAM. Image courtesy of NASA.

Polymers Erosion and Contamination Experiment (PEACE) and Polymers Experiments on MISSE 2, 5, 6, 7, and 8

Expeditions: 8–28, Ongoing
Leading Space Agency: NASA
Curriculum Grade Levels: 9–12 (secondary), Undergraduate (college, postsecondary)

Participating Country: United States
Participating State: Ohio

Number of 9–12 Students (secondary): 28
Number of Undergraduate Students (college, postsecondary): 2

Experiment Description: PEACE is a collaboration between Hathaway Brown students and engineers at NASA's Glenn Research Center. The experiment studies 41 polymer samples that are part of MISSE. Managed by NASA's Langley Research Center, MISSE is a collection of thousands of material samples and devices mounted on the outside of the space station. Researchers plan to test the samples for long-term durability in the harsh environment of space.

Description of Student Participation and Activities: Students at Hathaway Brown School for girls help with preflight research, flight sample fabrication, pre-flight and postflight characterization, data analysis, paper writing and presentations. They analyze polymers, long-chain molecular materials often used for spacecraft applications due to their light weight and flexibility. Their goal is to determine which polymers can withstand ultraviolet radiation and atomic oxygen in low Earth orbit.

Description of Teacher Participation and Activities: The teacher (Patty Hunt, Director of Research) provides a weekly research seminar at Hathaway Brown School and helps the students with logistics and science fair entrances.

Grace Yi, Student, Hathaway Brown School: "As technology continues to advance, it is vital that we have knowledge of the world and also what happens around it. Taking into account the countless number of satellites and communications that primarily depend on space-related functions, and on a larger scale, our society, which depends so heavily on this rapid communication, information garnered from MISSE experiments is invaluable. Data from the ISS MISSE experiments provides scientists with knowledge on how to construct spacecraft and other space applications that are a necessity in the world today, as well as the most suitable materials to build them with as to ensure the best functionality and longevity."

Additional Comments on Student/Teacher Involvement in the Investigation: The Hathaway Brown School has collaborated with Glenn Research Center on the MISSE PEACE experiments since 1998. The students work at Glenn Research Center on Friday afternoons after school and six weeks full time in the summer throughout their high school careers. Usually, there are three to six students on the team at a time. The students enter their research in national and international science competitions and have won more than $80,000 in scholarships and awards because of their MISSE research.

Education Lead: Kim K. de Groh, NASA Glenn Research Center, Cleveland, Ohio, USA

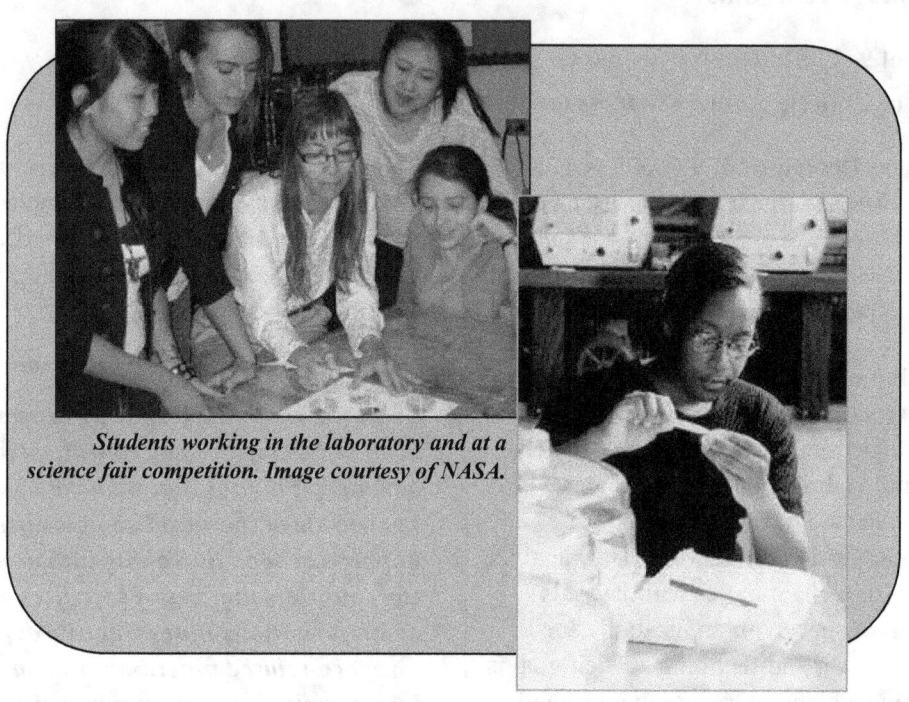

Students working in the laboratory and at a science fair competition. Image courtesy of NASA.

Robonaut

Expeditions: 27/28–33/34, Ongoing

Leading Space Agency: NASA

Curriculum Grade Levels: Undergraduate (college, postsecondary), Graduate (master's, Ph.D., M.D.)

Participating Country: United States

Number of K–8 Students (elementary): 300
Number of 9–12 Students (secondary): 300
Number of Undergraduate Students (college, postsecondary): 200
Number of Schools: 5
Number of Teachers: 12

Description of Student Participation and Activities: Student activities included question and answer sessions via Twitter — some of these were with one classroom, some were with multiple schools at once. Several members of the team visited various schools to conduct demonstrations. Live video conferences, which included students remotely controlling Robonaut, were held. Students visited the Robonaut laboratory for a live demonstration. The team also took Robonaut to the Smithsonian, where for an entire day classes passed through for live demonstrations and interactions with the Robonaut.

Education Lead: Joe Bibby, NASA JSC, Houston, Texas, USA
Education Website: http://robonaut.jsc.nasa.gov

Image of Robonaut courtesy of NASA.

Synchronized Position Hold, Engage, Reorient, Experimental Satellites (SPHERES)

Expeditions: 8–28, Ongoing

Leading Space Agency: NASA

Curriculum Grade Levels: Undergraduate (college, postsecondary), Graduate (master's, Ph.D., M.D.)

Impact: A total of 15 test sessions have been performed on ISS to date, starting with Expedition 13 and continuing through Expedition 18. The first few test sessions verified the functionality of many general utilities applicable to a variety of space missions. While continual improvements are being made to general algorithms on SPHERES, specialized algorithms are also being developed and tested to advance the areas of autonomous docking and formation flight.

Participating Country: United States

Participating State: Massachusetts

Number of Undergraduate Students (college, postsecondary): 30

Number of Graduate Students (master's, Ph.D., M.D.): 23

Experiment Description: SPHERES is a testbed for satellite formation flying — the theories and calculations that coordinate the motion of multiple bodies maneuvering in microgravity. To achieve this inside the station's cabin, bowling-ball-sized spheres perform various maneuvers (or protocols), with one to three spheres operating simultaneously. The SPHERES experiment tests relative attitude control and station-keeping between satellites, retargeting and image plane filling maneuvers, collision avoidance and fuel balancing algorithms, and an array of geometry estimators used in various missions.

View of the SPHERES satellites floating in the Destiny laboratory module as seen by the Expedition 14 crew. Flight engineer Thomas Reiter is visible in the background. NASA image ISS014E08025

SPHERES consists of three self-contained satellites, which are 18-sided polyhedrons that are 0.2 meter in diameter and weigh 3.5 kilograms. Each satellite contains an internal propulsion system, power, avionics, software, communications, and metrology subsystems. The propulsion system uses carbon dioxide (CO_2), which is expelled through the thrusters. SPHERES satellites are powered by

AA batteries. The metrology subsystem provides real-time position and attitude information. To simulate ground station-keeping, a laptop will be used to transmit navigational data and formation flying algorithms. Once these data are uploaded, the satellites will perform autonomously and hold the formation until a new command is given. SPHERES is an ongoing demonstration. Each session tests progressively more complex two and three-body maneuvers that include docking (to fixed, moving, tumbling targets), formation flying and searching for "lost" satellites.

On December 9, 2009, the Zero Robotics pilot teams observed the demonstration of their code running on the SPHERES robots aboard the space station. Image courtesy of MIT.

Description of Student Participation and Activities: The MIT Department of Aeronautics and Astronautics (Aero/Astro) has developed a new curriculum linking engineering fundamentals with real-world demands on engineers. Students gain engineering experience through all stages of the process, from concept to design, implementation and operations. The course features active group learning experiences in both the classroom and laboratory, rigorous assessment and evaluation. Junior, senior, and graduate students attending MIT have the opportunity to enroll in this three-semester course that focuses on preparing the students for engineering in the real world. Students design and build aerospace vehicle prototypes. These cutting-edge vehicles are then transferred to the Aero/Astro graduate program, where they are used in research. Student projects are complemented by internships in industry. During the first semester of the course, students develop requirements and conduct design trades to balance cost, complexity, risk, performance and safety. They also conduct full system reviews and, once completed, begin development of the highest risk subsystems and procure long-lead components. In the second semester, students refine the design, begin assembly, test system components and complete a critical design review. Students in the third semester complete assembly, conduct environmental and performance verification tests, conduct a hardware acceptance review and deploy the system.

Principal Investigator: David W. Miller, MIT, Cambridge, Massachusetts, USA
Education Website: http://ssl.mit.edu/spheres/

Educational Demonstrations and Activities

In addition to the numerous experiments students are supporting on ISS, several other educational activities and demonstrations are performed to be used as teaching aids, to supply educational resource materials, or simply to provide additional mechanisms to inspire students.

Butterflies, Spiders and Plants in Space

Expeditions: Ongoing

Leading Space Agency: NASA

Curriculum Grade Levels: K–8 (elementary), 9–12 (secondary)

Impact: These three projects involved more than 5000 teachers and 370,000 students worldwide. (Please note these numbers are accounted for in the CSI:03 and CSI:05 investigations.)

Participating Country: United States

Participating States: All

Number of K–12 Students (elementary): 370,000

Number of Teachers: 5000

Project Description: Four life science experiments conducted on the International Space Station from 2008 to 2012 demonstrated the effectiveness of using the ISS as a platform for student-centered experiments and STEM learning. The first experiment, which was a pilot, featured orb-weaver spiders (*Araneidae*) and painted lady butterflies (*Vanessa cardui*) transported inside special habitats on Space Shuttle Endeavour (STS-126). Once on board the ISS, the habitats were placed inside compartments that provided suitable living environments and permitted hourly photographs and video of the organisms' life cycles and behaviors in microgravity. Related teaching materials and photo archives were made available on the World Wide Web. The successful pilot led to three large-scale projects involving more than 5000 registered teachers and 370,000 students worldwide.

Butterflies in Space Teachers Guide.
Image courtesy of BioServe Space Technologies.

The 2009 experiment featured four painted lady butterfly larvae, which underwent metamorphosis and emerged as adults while in microgravity. In 2011, two golden orb spiders (*Nephilla clavipes*) and mustard seeds (*Brassica rapa*) were transported to ISS on Space Shuttle Endeavour (STS-134) for experiments conducted in the spring and fall of the same year. Student investigators replicated the ISS habitats in their classrooms and coordinated their observations with near-real-time downloads of images, videos and other data posted on Baylor College of Medicine's (BCM's) BioEd Online (www.bioedonline.org) website. Students collected and shared data and observations as they sought answers to their own questions related to life in microgravity. BioServe Space Technologies of the University of Colorado and BCM were partners in the educational experiments, which were conducted with support and collaboration from the NASA and the National Space Biomedical Research Institute. All project-related teaching resources and photo archives continue to be available free of charge for use by students and teachers around the world. Given the inexpensive materials required for these experiments, the open-ended nature of the student investigations, the availability of free curricular guides and resources, and the ongoing interest in space-based content, this model is highly replicable and likely to appeal to students and teachers from a wide range of primary and secondary schools.

Education Lead: Nancy Moreno, Ph.D., Baylor College of Medicine, Houston, Texas, USA
Education Website: http://bioedonline.org

Plants in Space.
Image courtesy of BioServe Space Technologies

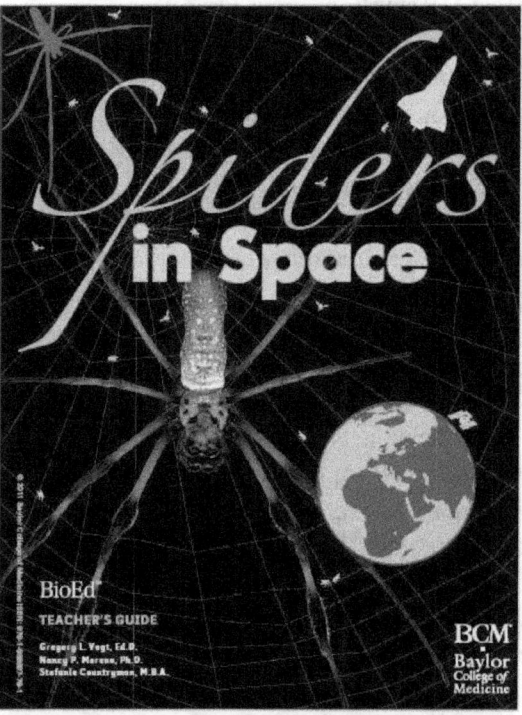

Spiders in Space.
Image courtesy of BioServe Space Technologies.

Education Payload Operations – Demonstrations (EPO-DEMOS)

Expeditions: 1–34, Ongoing

Leading Space Agency: NASA

Curriculum Grade Levels: K–8 (elementary), 9–12 (secondary)

Impact: More than 4 million students at the K–12 level have participated in education payload operations demonstrations. These demonstrations are used as teaching aids to inspire and motivate students in the study of science, mathematics and technology.

Participating Country: United States

Participating States: All

Number of K–8 Students (elementary): 2 million

Number of 9–12 Students (secondary): 2 million

Number of Teachers: 1.5 million

Project Description: Education Payload Operation – Demonstrations (EPO-Demos) records video education demonstrations performed on the International Space Station by crewmembers using hardware already onboard the ISS. EPO-Demos enhance existing NASA education resources and programs for educators and students in grades K–12. EPO-Demos support the NASA mission to inspire the next generation of explorers. The products are used for demonstrations and to support curriculum materials that are distributed across the United States and internationally to educators to encourage students to pursue studies and careers in science, technology, engineering and mathematics and inspire the next generation of space explorers. The individual EPO projects are designed to explore physical phenomena such as force, motion, and energy. Each International Space Station Expedition involves different on-orbit activities and themes as well as different partners, such as museums, universities, and public school districts.

Education Lead: Trinesha Dixon, NASA JSC, Houston, Texas, USA

Education Website: http://www.nasa.gov/education/tfs

Images of education demonstrations conducted by crewmembers on board the International Space Station. Images courtesy of NASA.

Table 12 – NASA Education Payload Operations – Demonstrations

No.	Demo	Objective	Expedition	Crewmembers
1	EPO-Tools-Demo	Compare ISS and EVA tools and their use to similar tools on Earth; discuss adaptations.	7	Edward (Ed) Lu
2	EPO-Newton's Laws-Demo	Use bodies to demonstrate Newton's Laws in microgravity.	7	Edward (Ed) Lu
3	EPO-Center of Mass-Demo	Demonstrate center of mass using rolled socks and string.	7	Edward (Ed) Lu
4	EPO-Lewis and Clark-Demo	Collaboration between CEO and Teaching from Space. Crew photographs significant sites on path of Lewis and Clark; still images.	7	Edward (Ed) Lu
5	EPO-Bernoulli's Principle-Demo	Demonstrate Bernoulli's Principle with paper and straw.	8	C. Michael (Mike) Foale, Alexander Kaleri
6	EPO-Lab Tour-Demo	Give a complete tour of the lab pointing out significant items.	8	C. Michael (Mike) Foale, Alexander Kaleri
7	EPO-Living Area-Demo	Show personal living space.	8	C. Michael (Mike) Foale, Alexander Kaleri
8	EPO-Hygiene-Demo	Demonstrate simple hygiene activities and compare to the same activities on Earth.	8	C. Michael (Mike) Foale Alexander Kaleri
9	EPO-Water Droplet-Demo	Show how a water droplet behaves in microgravity.	9	E. Michael (Mike) Fincke
10	EPO-Conservation of Angular Momentum-Demo	Spin body and move arms to show conservation of angular momentum.	9	E. Michael (Mike) Fincke
11	EPO-Weight and Mass-Demo	Demonstrate and explain the difference between weight and mass.	9	E. Michael (Mike) Fincke
12	EPO-Free Time-Demo	Demonstrate and discuss what the crew does during their free time.	12	William (Bill) McArthur
13	EPO-First Aid-Demo	Show medical supplies and discuss medical care on board.	12	William (Bill) McArthur
14	EPO-Sports-Demo	Demonstrate a variety of sports on orbit and use Newton's Laws to describe what happens.	12	William (Bill) McArthur
15	EPO-Sleep-Demo	Demonstrate and explain how the crew sleeps and eats; compare to similar activities on Earth.	12	William (Bill) McArthur
16	EPO-Arm Operations-Demo	Demonstrate robotic arm work stations and describe how they work.	12	William (Bill) McArthur

No.	Demo	Objective	Expedition	Crewmembers
17	EPO-Spacesuits-Demo	Show and compare Russian and American spacesuits.	12	William (Bill) McArthur Valery Tokarev
18	EPO-Supplies-Demo	Show a Progress and the supplies it brings.	12	William (Bill) McArthur
19	EPO-Recycling-Demo	Show and explain how water gets to the on the ISS and discuss the importance and conservation of resources.	12	William (Bill) McArthur
20	EPO-Solar Arrays-Demo	Explain how the solar panels work and show and describe uses for electricity.	12	William (Bill) McArthur
21	EPO-Lab Safety-Demo	Demonstrate lab safety procedures and compare to lab procedures on Earth.	12	William (Bill) McArthur
22	EPO-Floor/Ceiling-Demo	Show that in microgravity the crew can work in different planes.	12	William (Bill) McArthur
23	EPO-Vibrations-Demo	Demonstrate how daily activities produce vibrations and talk about problems associated with vibrations.	12	William (Bill) McArthur
24	EPO-Morning Routine-Demo/EPO-ECLSS-Demo	Describe the crewmembers' morning routine and Environmental Control and Life Support System.	14	Sunita (Suni) Williams, Michael (Mike) Lopez-Alegria
25	EPO-Fitness-Demo	Provide an overview of the exercise equipment used on ISS and the importance of fitness.	14	Sunita (Suni) Williams, Michael (Mike) Lopez-Alegria
26	EPO-Comm-Demo	Highlight the different equipment used on board the ISS for communicating with the ground.	14	Sunita (Suni) Williams, Michael (Mike) Lopez-Alegria
27	EPO-IPY-Demo	Describe ISS role in polar investigations and International Polar Year.	14	Sunita (Suni) Williams, Michael (Mike) Lopez-Alegria
28	EPO-Sports-Demo	Describe how sports would be different in microgravity.	15	Clayton (Clay) Anderson
29	EPO-Hobbies-Demo	Describe crewmember hobbies on the ISS.	15	Clayton (Clay) Anderson
30	EPO-Earth,Moon,Mars-Demo	Describe the distance and size between the Earth, moon, and Mars.	15	Clayton (Clay) Anderson
31	EPO-Sanitation on Station-Demo	Describe how crewmembers keep the ISS sanitary.	16	Peggy Whitson
32	EPO-ISS Living Area Tour-Demo	Describe the living area onboard the ISS.	16	Daniel (Dan) Tani
33	EPO-Newton's Laws/Rotation-Demo	Describe Newton's Laws and rotation in microgravity.	16	Daniel (Dan) Tani

No.	Demo	Objective	Expedition	Crewmembers
34	EPO-ISS Research Demo-Nutrition	Describe the nutrition research being conducted on the ISS.	16	Peggy Whitson
35	EPO-ISS Research Demo-InSPACE	Describe InSPACE research dealing with magnets that is conducted on the station.	16	Peggy Whitson
36	EPO-ISS Research Demo-SPHERES	Describe the SPHERES experiment and research being conducted on the ISS.	18	Gregory (Greg) Chamitoff, E. Michael (Mike) Fincke
37	EPO-Buzz Lightyear-Demo	Developed for Walt Disney. Not available for public distribution.	18	Gregory (Greg) Chamitoff, E. Michael (Mike) Fincke
38	EPO-Careers-Demo	Describe careers involved with ISS.	18	Sandra (Sandy) Magnus
39	EPO-ARISS Contact-Demo	Describe ham radio contacts from the ISS.	18	E. Michael (Mike) Fincke
40	EPO-ARISS-Demo	Demonstrate a live school contact.	18	Sandra (Sandy) Magnus
41	EPO-Renovation-Demo	Provide a tour of the ISS and describe the construction and renovations.	18	Sandra (Sandy) Magnus
42	EPO-Spacesuits2-Demo	Describe a spacesuit and what it is like to work inside of one.	18	Sandra (Sandy) Magnus
43	EPO-CEO-Demo	Describe CEO that take place on the ISS.	18	E. Michael (Mike) Fincke
44	EPO-Weight vs. Mass-Demo	Describe the difference between weight and mass using demonstrations in microgravity. Used with SSC Notice of Intent (NOI).	19/20	Robert (Bob) Thirsk, Koichi Wakata
45	EPO-SLAMMD-Demo	Describe the Space Linear Acceleration Mass Measurement Device experiment and how it applies to Newton's Second Law of Motion: F=MA.	19/20	Michael (Mike) Barratt, Koichi Wakata
46	EPO-Recycling-Demo	Describe how astronauts recycle their urine and sweat into drinking water. Used with Waste Limitation, Management, and Recycling (WLMR) Kennedy Space Center NOI.	19/20	Robert (Bob) Thirsk, Timothy (Tim) Kopra
47	EPO-Eating-Demo	Describe how astronauts eat in space.	19/20	Robert (Bob) Thirsk, Frank DeWinne
48	EPO-Bernoulli's Principle-Demo	Describe Bernoulli's Principle using demonstrations in microgravity.	19/20	Koichi Wakata, Frank DeWinne
49	EPO-ISS Research Demo-LOCAD	Describe the Lab-on-a-Chip Application Development (LOCAD) experiment.	19/20	Koichi Wakata, Michael (Mike) Barratt

No.	Demo	Objective	Expedition	Crewmembers
50	EPO-Linear Momentum-Demo	Describe linear momentum using demonstrations in microgravity.	19/20	Robert (Bob) Thirsk, Frank DeWinne
51	EPO-Time-Demo	Describe time zones and what time it is on the ISS.	19/20	Robert (Bob) Thirsk, Frank DeWinne
52	EPO-Weight vs Mass2-Demo	Describe the difference between weight and mass using demonstrations in microgravity. Used with SSC NOI.	19/20	Robert (Bob) Thirsk, Nicole Stott
53	EPO-Wave Motion-Demo	Describe waves, their motion, and interactions.	19/20	Robert (Bob) Thirsk, Frank DeWinne
54	EPO-Surface Tension-Demo	Describe surface tension of water and intermolecular forces.	19/20	Robert (Bob) Thirsk, Frank DeWinne
55	CBPD	Continuous Blood Pressure Device heart experiment	19/20	Robert (Bob) Thirsk
56	IRIS Experiment	Eye experiment	19/20	Robert (Bob) Thirsk
57	Neurospat	Brain and Nervous System	19/20	Robert (Bob) Thirsk
58	BISE	Bodies in the Space Environment experiment	19/20	Robert (Bob) Thirsk
59	Integrated Immune	Immune System Experiment	19/20	Robert (Bob) Thirsk
60	CCISS	Blood Pressure Experiment	19/20	Robert (Bob) Thirsk
61	EPO-Rotation Set-Demo	Describe center of mass, rotation, and moments of inertia.	21/22	Jeffrey (Jeff) Williams
62	EPO-Speed/Work/Energy-Demo	Describe differences in speed between ISS and celestial objects. Describe the work and energy involved.	21/22	Jeffrey (Jeff) Williams
63	EPO-Eye in the Sky-Demo	Describe the many phenomena that are uniquely observable from the ISS.	21/22	Jeffrey (Jeff) Williams
64	EPO-Centripetal Force/Acceleration-Demo	Describe centripetal acceleration, how it arises, and its practical uses onboard the ISS.	21/22	Jeffrey (Jeff) Williams
65	EPO-Sesame Street 1-Demo	This video is developed in collaboration with Sesame Street and will be used in Season 41. Letter A/F is discussed and Word on the Street-Float.	22/23	Soichi Noguchi
66	EPO-Cloud Observation-Demo	This video is part of the S'COOL collaboration and discusses cloud formations as viewed from the ISS.	23/24	Tracy Caldwell-Dyson
67	EPO-Dress4Scess-Demo	Describe why astronauts dress the way they do for work and why clothing is made differently for space.	25	Douglas (Doug) Wheelock

No.	Demo	Objective	Expedition	Crewmembers
68	EPO-Exercise2-Demo	Describe how astronauts exercise in space.	25	Douglas (Doug) Wheelock
69	EPO-Hobbies2-Demo	Describe what hobbies the astronauts take part in on the ISS and how those hobbies are impacted by the microgravity environment.	25	Douglas (Doug) Wheelock
70	EPO-Sesame Street 2-Demo	This video is developed in collaboration with Sesame Street and will be used in Season 42. Counting to 10 using floating items.	27	Catherine (Cady) Coleman, Ronald (Ron) Garan
71	EPO-EarthMoonMars-Demo	Describe the differences in distance and size between the Earth, moon, and Mars. Describe scale modeling.	29	Michael (Mike) Fossum
72	EPO-Gyroscopes-Demo	Describe how gyroscopes stabilize the ISS.	29	Michael (Mike) Fossum
73	EPO-Sleep2-Demo	Describe the crew sleeping quarters including those in the Russian segment.	29	Satoshi Furakawa, Michael (Mike) Fossum, Sergei Volkov
74	EPO-Eye in the Sky 2-Demo	Illustrate the many phenomena that are uniquely observable from the Cupola.	29	Michael (Mike) Fossum, Satoshi Furukawa
75	EPO-R2EDU101-Demo	Describe the purpose of robotics and Robonaut on the ISS.	29	Michael (Mike) Fossum
76	EPO-ISS Orbit-Demo	Describe and illustrate the orbit of the ISS. Describe the path of the ISS as illustrated on a world map.	29	Michael (Mike) Fossum
77	EPO-Space Sports-Demo	Demonstrate student designed games in the microgravity environment of the ISS.	30	Daniel (Dan) Burbank

From Jason Pittman, science teacher at Hollin Meadows Math and Science Elementary School in Alexandria, Virginia: "I was really impressed with the ideas that they came up with. As a teacher, I was pleasantly surprised at how thorough their investigation was. This is a great way for me to connect our state curriculum with something that really does get to kids' imaginations. The enthusiasm for science and for space and for NASA has all really, really just given kids a real charge about science in general. When they come into the science lab, they are just really interested."

From Lino, a student at Hollin Meadows Math and Science Elementary School in Alexandria, Virginia: "Most of us thought that they (the space seeds) would never grow, but we were wrong. They grew better than the Earth plants. Looking at the seeds under the microscope made them look like they were from another dimension. I didn't think they would look any different."

JAXA Education Payload Operations – Demonstrations (JAXA EPO–Demos)

Expeditions: 17/18, 19/20, 21/22, 23/24, 33/34
Leading Space Agency: Japan Aerospace Exploration Agency (JAXA)
Curriculum Grade Levels: K–8 (elementary), 9–12 (secondary)

Participating Country: Japan

Description of Public Participation and Activities: The JAXA EPO-Demos allow the Japanese community the opportunity to interact with International Space Station crewmembers through various activities for educational purposes. These activities help to enlighten the general public about microgravity utilization and human space flight via crewmember performances of on-orbit demonstrations exhibiting that microgravity is useful not only for scientists and engineers, but also for writers, poets, teachers, artists and others.

Description of Student Participation and Activities: The Education Payloads Operation for education consists of various activities that attract attention to Japanese manned space flight activities to gain the support of the Japanese community for future manned space exploration.

Principal Investigator: Naoko Matsuo, Japan Aerospace Exploration Agency, Tsukuba, Japan

JAXA astronaut Satoshi Furukawa, Expedition 29 flight engineer, demonstrates the building of a model space station in the Kibo laboratory of the International Space Station. Image ISS029e011045 (October 2011)

JAXA astronaut Soichi Noguchi, Expedition 23 flight engineer, watches a water bubble float freely between him and the camera, showing his image refracted, in the Kibo laboratory of the International Space Station. Image ISS023-E-025091 (April 19, 2010)

Table 13 – JAXA Education Payload Operations – Demonstrations

No.	Demo	Objective	Expedition	Crewmembers
1	JAXA EPO 1	Modeling Clay in Space shows the inspiration of space and the human space flight by clay figures. Videos are taken by the crewmember to document the experiment. The dried figures are returned to Earth and used as art for a tribute to space flight.	17, 18, 19/20, 21/22, 23/24	Koichi Wataka, Soichi Noguchi
		Artistic Experiments Using a Water Sphere aims to record the motion of a water sphere after external oscillations in microgravity environment with a high-definition television (HDTV) camera.		
		Marbling Painting on a Water Ball records and obtains the ink flow on the water sphere.		
		Sparkling Neurons will obtain imagery with HDTV. Imagery will be returned to Earth for examination. The investigator will then create an intuitive image of the space environment.		
		"Moon" score: A space station astronaut will capture various pictures of the Moon from the ISS. Images will be returned to Earth, where a musical score will be composed to match the images.		
2	JAXA EPO 2	Spiral Top aims to record the motion of a luminous spinning top onboard the ISS.	18, 19/20, 21/22,	Koichi Wataka, Soichi Noguchi
		Space Clothes investigates body movement in microgravity and obtains basic data for future clothes.		
		Hiten (Dance) records a crewmember performing some postures of ancient east Asian flying deities in microgravity.		
		Space Poems Chain consists of a collection of poems composed by famous poets and the general public, which is recorded on a Digital Video Disk (DVD) and stored on board ISS in the Japanese Experiment Module (JEM).		
3	JAXA EPO 3	Dewey's Forest showed how gravity controls the laws of nature and influences our ways of thinking. The project was a catalyst to rediscover our relationship with plants on the ground and the age old history of our gardens.	19/20, 21/22, 23/24	Koichi Wataka, Soichi Noguchi
4	JAXA EPO 4	Crewmembers videotaped or took digital imagery of JAXA EPO4 activities in the ISS/JEM.	23/24	Soichi Noguchi

No.	Demo	Objective	Expedition	Crewmembers
5	JAXA EPO 5	Message in a Bottle created a unique communication interface between space and Earth, as well as present and future humankind. During an EVA, a small cylinder was filled with space. The astronaut created a memento of his or her time in space, but also a message for people on Earth. Once the "Message in a Bottle" was brought back to Earth and placed in people's hands, it became a conduit between humans and space, and between this world and the one beyond us.	23/24	Soichi Noguchi
6	JAXA EPO 6	Spiral Top-II aims to record the motion of a luminous spinning top on board the ISS.	23/24	Soichi Noguchi
7	JAXA EPO 7	These were artistic experiments and cultural activities. These activities were implemented to enlighten the general public about microgravity utilization and human space flight. JAXA understands that the space station's JEM, Kibo, is useful for scientists and engineers as well as writers, poets, teachers, artists and others.	27/28, 29/30	Satoshi Furukawa
8	JAXA EPO 8	This is a demonstration of educational events and artistic performances on board the Kibo module. These activities are expected to enlighten the general public about the wonders of microgravity phenomena and human space flight.	29/30	Satoshi Furukawa
9	JAXA EPO 9	EDUCON consists of educational events and artistic performances on board the Kibo module through the EPO 9 activities. These activities are expected to enlighten the general public about the wonders of microgravity phenomena and human space flight. JAXA REPORT is an activity to write a report in Japanese concerning ordinary life on the station to attract attention for manned space activity and to gain the support of the Japanese public for future manned space exploration. SPACE SOUND: We will investigate the effects of the space environment on the feelings, opinions and minds of astronauts. We will focus our study on humans' sense of hearing. Astronauts will use a bowl made of a copper alloy called CHUON, a sound emitting object, to produce a sound and listen to it.	31/32	Akihiko Hoshide

No.	Demo	Objective	Expedition	Crewmembers
		UNWINDING: We will search for conditions that promote relaxed living, utilizing the entirety of a limited space, with no gravity, by exploring how people's sense of space in the ISS and on the ground differs. In doing so, we will contribute to the design of comfortable living spaces in the ISS and, in turn, come up with new ideas for the design of living spaces on the ground.		
		VIDEO: This activity is to shoot a tour of the station, introducing Kibo experiments and life in space, to show life in space to children and the public to gain support for manned space exploration. The appearance of these experiments is recorded with the G1 camcorder and downlinked for the purpose of scientific study.		
		REPORT is an activity to write a report in Japanese concerning ordinary life on the station to attract attention for manned space activity and to gain the support of the Japanese public for future manned space exploration.		
10	JAXA EPO 10	EDU CON: This activity is to shoot the educational activity demonstration in space. These experiment appearances are recorded with the G1 camcorder and downlinked for the purpose of scientific study.	31/32	Akihiko Hoshide
		VIDEO: This activity is to shoot a tour of the station, introducing Kibo experiments and life in space, to show life in space to children and the public to gain support for manned space exploration.		
		REPORT is an activity to write report in Japanese concerning ordinary life on the station to attract attention for manned space activity and to gain the support of the Japanese public for future manned space exploration.		

ESA – Mission Digital Video Disk (DVD) Series

Expeditions: 7, 9, 11, 13, 14
Leading Space Agency: ESA
Curriculum Grade Level: 9–12 (secondary)
Impact: Twenty thousand prints of each DVD (1–4) have been printed and distributed to European schools. The DVD series is also available on YouTube and has several thousand views months after publication. The DVD series is an ideal, modern and fast paced modular film series that can easily be used by teachers to suit their teaching needs. The DVDs cover topics of Newton's Laws, the effect of space on the body, chemistry of matter and robotics.

Participating Countries: Austria, Belgium, Czech Republic, Denmark, Finland, France, Germany, Greece, Ireland, Italy, Luxembourg, Netherlands, Norway, Portugal, Romania, Spain, Sweden, Switzerland, United Kingdom

Number of Secondary School Students: >10,000
Number of Teachers: >500
Number of Schools: >500

Description of Student Participation and Activities: The DVD series is an ideal, modern and fast-paced modular film which can be used easily by teachers to suit their teaching needs. The DVDs cover topics of Newton's Laws, the effect of space on the body, chemistry of matter and robotics. During their missions, ESA astronauts Pedro Duque, André Kuipers, Roberto Vittori and Thomas Reiter participated in this DVD series by performing experiments on board the ISS. Assisted by students on the ground, all aspects of human space flight and space in general are touched upon in this series.

Schools can order the educational kits by clicking the link http://wsn.spaceflight.esa.int/education/.

Education Lead: Nigel Savage, Ph.D., ESTEC, Netherlands
Education Website: www.esa.int/spaceflight

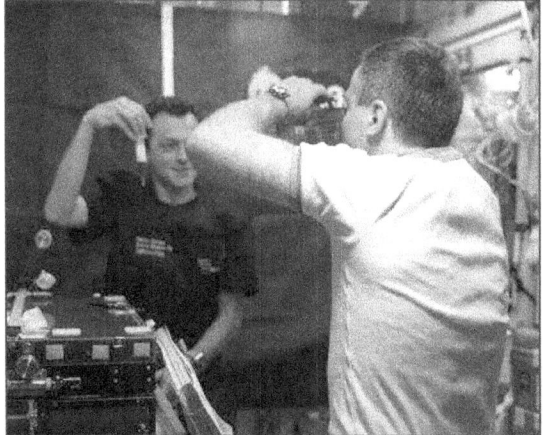

Pedro Duque is filmed by Alexander Kaleri for the educational experiment which will demonstrate basic physical phenomena featured in Mission 1 DVD: Newton in Space. Image courtesy of ESA.

Table 14 – ESA DVD and Educational Kits

Demo	Objective	Expedition	Crewmember
ISS-DVD-1 – Newton in Space	This demonstration explains Newton's Three Laws of Motion.	7	Pedro Duque
ISS-DVD-2 – Body in Space	This demonstration shows a number of physiology demonstrations showing the effects of weightlessness on the human body.	9	André Kuipers
ISS-DVD-3 – Space Matters	This demonstration was designed to explore the different structures, states and properties of matter.	11	Roberto Vittori
ISS-DVD-4 – Space Robotics	This demonstration includes a series on space robotics.	13, 14	Thomas Reiter
Kit – Primary	ESA has produced a binder for primary school teachers who are eager to use space as a motivating tool for teaching subjects ranging from science to art. The A4-sized binder contains four chapters devoted to explaining various aspects of life in space and what is it like to live and work on board the International Space Station. Each chapter contains background information, worksheets and a teacher's guide. A glossary and color posters complete the content. The kit can be used as the basis for classroom lessons or group exercises or be given out as homework. Individual pages can be photocopied and distributed to each pupil. The kits can be downloaded in various languages in pdf format.		
Kit – Secondary	In 2001, ESA organized Teach Space 2001, our first International Space Station Education Conference, providing an opportunity for teachers to exchange practical ideas on the theme of space. The recommendations of the conference led to the production of a pilot ISS Education Kit for use by teachers of pupils aged 12 to 15 years old (International Standard Classification for Education – ISCED – level 2). Five chapters explain various aspects of the International Space Station, including what it is, how it is being built, living and working on board, and what future voyages will be like. There are not only text and explanations about the International Space Station, but related inter-disciplinary exercises, a teacher's guide, a glossary and color overhead transparencies. Modules can be taught in the classroom, used to facilitate group exercises, given as homework, or copied and distributed to each pupil.		

ESA – Space in Bytes Series

Expeditions: 16, 27/28
Leading Space Agency: ESA, Global
Impact: More than 85,000 views

Ongoing Possibilities: The Space in Bytes series is still being implemented, and new clips will be published with each expedition.

Description: ESA is developing a series of short video lessons (Space-in-Bytes) for upper secondary school students and their teachers. The extensive use of videos and the Internet by youngsters today makes Space-in-Bytes an attractive and innovative tool to reach out to a broad audience of students.

Space in Bytes, ESA's short educational films.

The aim of Space-in-Bytes is to develop educational content in the form of online video lessons that
- Explain the importance of space research
- Show and explain scientific phenomena that have a different behavior in the space environment
- Show and explain scientific phenomena that are relevant to the European science curricula
- Familiarize pupils with the uniqueness of the weightless environment
- Increase the interest of upper secondary school students in science and technology.

Space-in-Bytes is based on key space topics related to the International Space Station, future space exploration and current ESA research projects. Each video is accompanied by an online lesson that includes scientific explanations of some of the concepts illustrated in the video. The lesson is intended to provide teachers and their students more information about the topics covered in the video and suggest curriculum-relevant exercises that can be carried out in class.

Space-in-Bytes is intended to present "byte-size" scientific information, offering a starting point for further individual or classroom investigation. These videos are available to download from ESA's website in several languages and are also available from ESA's YouTube educational playlists in most ESA member state languages.

Education Lead: Nigel Savage, Ph.D., ESTEC, Netherlands
Website: http://www.esa.int/SPECIALS/Space_In_Bytes/index.html

Global Water Experiment

Leading Space Agency: NASA, Global

Curriculum Grade Levels: K–8 (elementary), 9–12 (secondary)

Participating Countries: Global

Participating States: California, Florida, Georgia, Iowa, Indiana, Massachusetts, Maryland, Michigan, Minnesota, Oregon, Pennsylvania, Texas, Utah, Virginia, Wisconsin

Number of K–8 (elementary), 9–12 Students (secondary): More than 24,000

Number of Schools: 619

Number of Teachers: 1174

Students at Pacifica High School conduct a water filtration experiment. Image courtesy of Pacifica High School.

Description of Student Participation and Activities: The United Nations Educational, Scientific and Cultural Organization (UNESCO), in partnership with the International Union of Pure and Applied Chemistry (IUPAC), designated 2011 as the International Year of Chemistry. As a part of the year-long activities, students across the world sample water in their communities and conduct experiments to learn about the pH of the planet, the salinity content in water, various water treatment and disinfection methods and methods of desalination. The results collected will be used to build an interactive, global water data map.

The American Chemistry Society is sponsoring Water: A Chemical Solution that will provide students in the United States an opportunity not only to learn about the water challenges other people face but to encourage interest in chemistry among young people, generate enthusiasm for the creative future of chemistry and enhance international cooperation. Through this experience, students learn how the International Space Station water resources are managed. They learn about the fundamental technology that is used as a part of the Environmental Control Life Support System, which handles up to 23.2 pounds of condensate, crew urine, and urinal flush water to produce a purified distillate. This distillate is combined with other wastewater sources collected from the crew and cabin and is processed, in turn, by a water processor assembly (WPA) to produce drinking water for the crew. The ISS WPA has an iodinated resin that is used in the filtration process. Iodine is added to the water to control the growth of microorganisms — just like chlorine is added to the water we drink at home. However, iodine is used instead of chlorine because iodine is much easier to transport to orbit and it is less corrosive. This iodinated resin has been developed as a commercial water filtration solution for use in disaster and humanitarian relief zones in a number of countries throughout the world.

Education Lead: Matthew Keil, Teaching From Space, NASA JSC, Houston Texas, USA

Education Websites: http://www.nasa.gov/education/tfs, http://water.chemistry2011.org

Rita Nobile, 15, of Pacifica High School, speaks to the members of NASA's Digital Learning Network as part of a webcast. Photo by Anthony Plascencia, Ventura County Star.

International Space Station In-flight Education Downlinks (In-flight Education Downlinks)

Expeditions: 1–34, Ongoing

Leading Space Agency: NASA, CSA, ESA, JAXA, Roscosmos

Curriculum Grade Levels: K–8 (elementary); 9–12 (secondary)

Impact: More than 37 million students ranging from elementary to secondary school levels have participated in in-flight live downlinks from the station.

Participating Countries: Belgium, Canada, England, Greece, Italy, Japan, Russia, Spain, United States

Participating States: Alabama, Alaska, Arizona, California, Connecticut, Florida, Georgia, Hawaii, Idaho, Illinois, Indiana, Iowa, Kansas, Kentucky, Maine, Maryland, Massachusetts, Michigan, Minnesota, Mississippi, Missouri, Montana, Nebraska, Nevada, New Mexico, New York, North Carolina, North Dakota, Ohio, Oklahoma, Oregon, Pennsylvania, Rhode Island, South Carolina, Tennessee, Texas, Utah, Virginia, Washington, West Virginia, Wisconsin

Number of K–8 Students (elementary): 24,870,924

Number of 9–12 Students (secondary): 12,436,392

Number of Schools: 825

Number of Teachers: 1,285,646

Description of Student Participation and Activities: Downlinks afford students the opportunity to learn first-hand from space explorers what it is like to live and work in space. Students participate in educational activities that prepare them to pose questions to the on-orbit crew. During a downlink, 20 students have the opportunity to ask the crew questions, and the audience watches as crewmembers discuss and demonstrate STEM concepts in ways that are unique to the environment of space. Following the downlink, students participate in educational activities to continue the impact of the downlink.

Expedition 13 crewmembers on board the space station answered questions students asked in four languages during the International Education Week event held November 14, 2006, at the U.S. Department of Education. Image courtesy of NASA Educational Technology Services.

Description of Teacher Participation and Activities: This activity is mainly student focused, but teachers actively prepare the students for a downlink by conducting preliminary activities designed to increase the students' knowledge of space, the astronauts and a variety of STEM topics. Additionally, teachers also conduct educational activities after the downlink to extend the impact of the downlink.

Education Leads: Becky Kamas, NASA JSC; Naoko Matsuo, JAXA; Nigel Savage, Ph.D., ESA
Education Website: http://www.nasa.gov/education/tfs

NASA Associate Administrator for Education Leland Melvin and NASA education specialist Trinesha Dixon with a New York high school student (center) await a response from astronaut Catherine (Cady) Coleman aboard the International Space Station during a special downlink held in honor of Women's History Month. (Women's Academy of Excellence, New York, New York.) Image courtesy of NASA/Bill Ingalls.

From Lori Beth Bradner, science teacher at Central Florida Aerospace Academy in Lakeland, Florida, USA: "A lot of our students at the Central Florida Aerospace Academy, because of our focus, they want to be those people. So, for them to talk to those leaders, the people that are living the dream they have, was unbelievable."

From Alejandro Aybar-Moto, student at Central Florida Aerospace Academy in Lakeland, Florida, USA: "That's a moment I'll never forget. It's going to live with me for the rest of my life."

From Colleen Brannan, Senior Assistant to the President of State University of New York – Oneonta in Oneonta, New York: "We're excited to bring so many students here for what's shaping up to be a fantastic experience. The college is thrilled to give these young people the opportunity to talk with an astronaut in orbit. We're also grateful to our alumnus, Colonel Garan for his enthusiasm and generosity in sharing his knowledge with kids from local schools."

International Space Station Ham Radio (ISS Ham Radio) [Also Known as Amateur Radio on the International Space Station (ARISS)]

Expeditions: 1–34, Ongoing

Leading Space Agency: NASA, CSA, ESA, JAXA, Roscosmos

Curriculum Grade Levels: K–8 (elementary), 9–12 (secondary)

Impact: Almost 120,000 students from 693 schools from 44 countries worldwide have participated.

Participating Countries: Argentina, Australia, Belgium, Brazil, Canada, Croatia, Ecuador, Finland, France, Germany, Greece, Hungary, India, Ireland, Israel, Italy, Japan, Kuwait, Luxembourg, Malaysia, Mali, New Zealand, Northern Ireland, Norway, Puerto Rico, China, Peru, Poland, Portugal, Republic of Korea, Russia, Senegal, Slovenia, South Africa, Spain, Sweden, Switzerland, Taiwan, Thailand, Netherlands, Trinidad and Tobago, Turkey, United Kingdom, United States

Participating States: Alabama, Alaska, Arizona, California, Colorado, Florida, Georgia, Hawaii, Idaho, Illinois, Indiana, Iowa, Kentucky, Maine, Maryland, Massachusetts, Michigan, Mississippi, Missouri, Montana, Nebraska, Nevada, New Hampshire, New Jersey, New Mexico, New York, North Carolina, North Dakota, Ohio, Oklahoma, Oregon, Pennsylvania, Rhode Island, South Carolina, South Dakota, Tennessee, Texas, Utah, Vermont, Virginia, Washington, West Virginia, Wisconsin, Wyoming and the District of Columbia and Puerto Rico

Number of K–8 Students (elementary): 82,000

Number of 9–12 Students (secondary): 37,000

Number of Schools: 693

Number of Teachers: 9500

"The experience has excited me in the field of astronomy by re-inspiring me to accomplish my dream to become an astronaut."

"The space communication experience has been truly remarkable; the ARISS program has been educational and influential on a career based on this field."

"The communication with the astronauts was an amazing experience. It was something I have never experienced before and I think that it is a program that should be continued and further improved. My interest is in accounts and business but this historical moment has made me begin to consider space a bit more. Thank you for the tremendous opportunity of being a part of this moment."

"I was a student who asked a question via radio to an astronaut aboard the ISS. It was a unique, informative, and most of all, an awesome experience to be able to establish contact."

Astronaut Sunita L. Williams, flight engineer on Expeditions 14 and 15, talks with students at the International School of Brussels in Belgium during an ARISS session in the Zvezda Service Module.
NASA image ISS014E18307

Description of Student Participation and Activities: International students and teachers use ham radio to talk with ISS crewmembers. In preparation for the contact, students research topics such as the ISS, space exploration, radio waves and amateur radio. They participate in hands-on activities such as building model rockets, models of the solar system or crystal radios, and they may listen to other amateur radio satellites. They use what they have learned to prepare a list of questions for the ISS crewmember about life in space, the ISS and the experiments onboard the ISS, or other space-related topics. Through this experience, students are inspired to continue their education in STEM fields and pursue STEM-related careers.

Description of Teacher Participation and Activities: Teachers prepare students for an ARISS contact through their science and math curricula, employing lesson plans that concentrate on such topics as force and motion, satellite orbits, satellite tracking, radio signals and wave propagation. A variety of NASA and Amateur Radio Relay League education resources are used in conjunction with these lessons.

Education Leads: Trinesha Dixon, NASA JSC; Naoko Matsuo, JAXA
Education Website: http://www.arrl.org/ARISS/

Students attending Space Camp at the Euro Space Center in Belgium are gathered in an auditorium to speak with astronaut Ed Lu, who was on board the station during Expedition 7 in July 2003. Image courtesy of ESA.

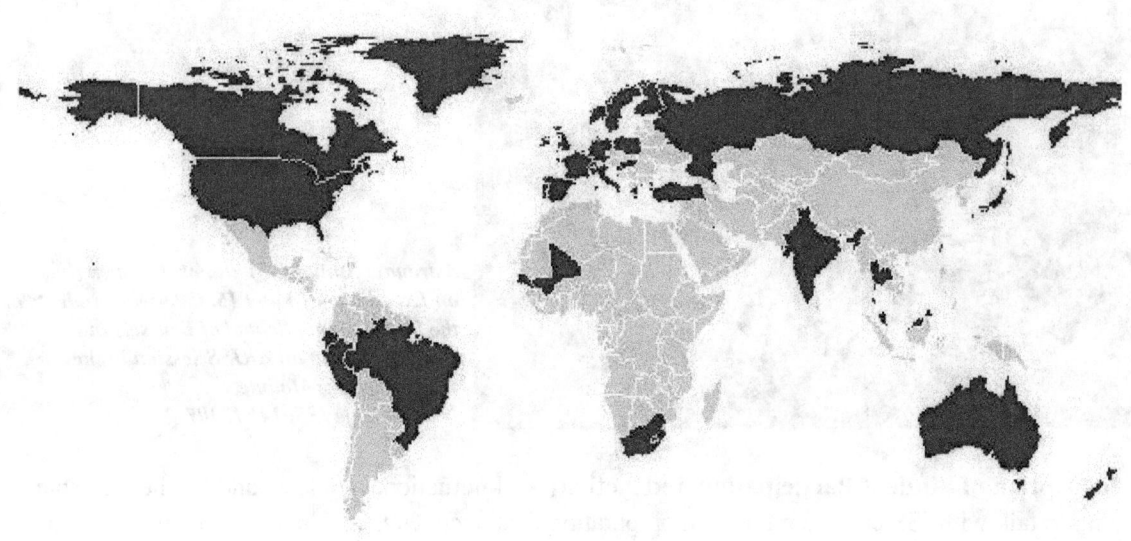

Map of countries that participated in ARISS missions.

Table 15 – Countries and Number of Schools That Conducted ARISS Contacts

Country	No. of Schools	Country	No. of Schools
Argentina	1	New Zealand	4
Australia	37	Northern Ireland	1
Belgium	33	Norway	2
Brazil	4	P.R. China	3
Canada	50	Peru	4
Croatia	1	Poland	7
Ecuador	1	Portugal	5
Finland	3	Republic of Korea	2
France	17	Russia	35
Germany	16	Senegal	1
Greece	3	Slovenia	1
Hungary	2	South Africa	4
Ile de La Réunion	1	Spain	6
India	6	Sweden	2
Ireland	1	Switzerland	9
Israel	2	Taiwan	1
Italy	76	Thailand	2
Japan	59	Netherlands	6
Kuwait	2	Trinidad and Tobago	1
Luxembourg	1	Turkey	2
Malaysia	9	United Kingdom	14
Mali	1	United States	252
Total – 693 schools			

Table 16 – ARISS Contacts by U.S. States

State	No. of Schools	State	No. of Schools
Alabama	2	Montana	2
Alaska	1	Nebraska	2
Arizona	4	Nevada	1
Arkansas	0	New Hampshire	1
California	20	New Jersey	8
Colorado	5	New Mexico	3
Connecticut	0	New York	25
Delaware	0	North Carolina	4
District of Columbia	8	North Dakota	0
Florida	13	Ohio	12
Georgia	2	Oklahoma	3
Hawaii	3	Oregon	3
Idaho	3	Pennsylvania	7
Illinois	18	Puerto Rico	2
Indiana	10	Rhode Island	0
Iowa	1	South Carolina	3
Kansas	0	South Dakota	0
Kentucky	5	Tennessee	2
Louisiana	1	Texas	39
Maine	1	Utah	2
Maryland	4	Vermont	0
Massachusetts	4	Virginia	8
Michigan	9	Washington	4
Minnesota	0	West Virginia	0
Mississippi	2	Wisconsin	1
Missouri	4	Wyoming	0
Total U.S. Schools – 252			

International Space Station *Live!* (ISS*Live!*)

Expeditions: 29/30, Ongoing

Leading Space Agency: NASA

Curriculum Grade Levels: K–8 (elementary), 9–12 (secondary), Undergraduate (college, postsecondary)

Participating Country: United States

Participating States: All

Description of Student Participation and Activities: Students participated in the beta testing of the website after it was released in October 2011 by accessing the oxygen generator lesson and interacting with the live data that was released. Students were exposed to the website. A contributing ISS flight controller gave a presentation at the University of Minnesota and informed the physics department about ISS*Live!*; the link to the ISS*Live!* website and a direct link to the ethos console handbook part of the page were added to the university's website to facilitate educational use of the oxygen generator lesson.

Description of Teacher Participation and Activities: Teachers participated in the beta testing of the website after it was released in October 2011 by accessing the oxygen generator lesson and interacting with the live data that was available. Teachers were exposed to the website.

Additional Comments on Student/Teacher Involvement in the Investigation: Many of our beta test viewers informed their educational friends (teachers from all over the country) about the ISS*Live!* website, but data on how many persons have proactively used the site in their classroom is unavailable (at the time of this publication). Mechanisms and processes are being pursued to obtain this information in 2012.

ISS*Live!* is available as a mobile application on the Android and iPhone.

Seven additional lessons that will use live data from the ISS and apply to additional console positions in the mission control room will be available for classroom use at the end of the 2012 school year.

Education Leads: Jennifer Price, NASA JSC; Dr. Regina B. Blue, NASA JSC; and Monica Trevathan, Tietronix

Education Website: http://spacestationlive.jsc.nasa.gov

ISSLive! Interact in 3D: Explore Mission Control. Images courtesy of NASA.

International Toys in Space

Expeditions: 5, Ongoing

Leading Space Agency: NASA, Global

Curriculum Grade Levels: K–8 (elementary), 9–12 (secondary)

Impact: To date, more than 500 videos, DVDs, and video clips have been produced and distributed to science teachers and schools throughout the United States. Each year, approximately 1500 teachers are trained to use the materials in their classrooms.

Participating Countries: Worldwide

Participating States: All

Description of Student Participation and Activities: The objective of the Education Payload Operations – International Toys in Space (EPO-International Toys in Space) investigation is to use toys, tools and other common items in the microgravity environment of the International Space Station to create educational video and multimedia products that inspire the next generation of engineers, mathematicians, physicists and other scientists. The products are used in demonstrations and to support curriculum materials distributed across the United States and internationally.

EPO-International Toys in Space involves students in the principles of physics by studying how everyday items (toys and games) act in a microgravity environment. A variety of toys are flown to the station, including international toys representing the International Space Station International Partners, and video is captured of crewmembers experimenting with the toys. The video footage is used to produce a DVD and an accompanying teacher's guide. Following the activity guide, students investigate how the toys behave on Earth and compare their results to video footage taken on board ISS. Both the video and the toys are available for teachers at NASA's Central Operation of Resources for Educators (CORE).

Education Lead: Matthew Keil, NASA JSC, Houston, Texas, USA

Education Website:
http://www.nasa.gov/education/tfs

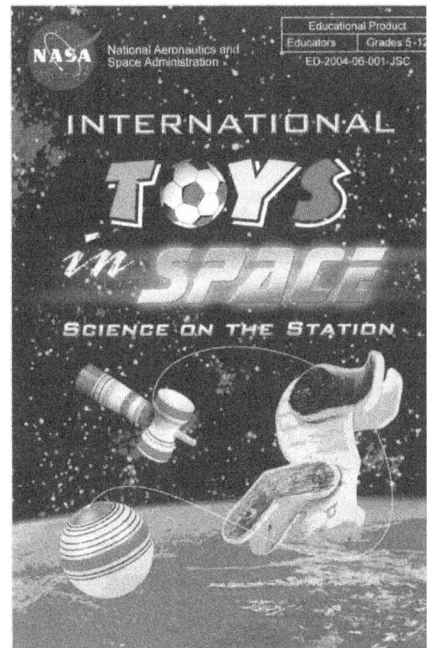

Cover of the DVD produced by the Education Payloads Office for Increment 5 EPO-International Toys in Space. This DVD is available to the public to help educate students about how toys are used differently in space than on Earth. Image courtesy of NASA.

Lego® Bricks

Expeditions: 27/28, Ongoing
Leading Space Agency: NASA, Global
Curriculum Grade Levels: K–8 (elementary), 9–12 (secondary)

Participating Countries: Worldwide
Participating States: All

Description of Student Participation and Activities: NASA has partnered with the LEGO Group to develop innovative educational materials and activities. This partnership, documented in a Space Act Agreement between the two organizations, is designed to support NASA's educational programs in exploration, technology, science and aeronautics. This opportunity provides the unique learning environment of microgravity to promote student interest in STEM content and careers. To accomplish this task, LEGO kits are flown on board the International Space Station. Crewmembers perform tasks to demonstrate simple science concepts and show how LEGO bricks work differently in a microgravity environment.

Students dig into a pile of LEGO® bricks to begin constructing their miniaturized visions of the future of space travel. Image courtesy of Jack Pfaller/NASA.

Students used LEGOs to "Build the Future" at NASA's Kennedy Space Center in Cape Canaveral, Florida. The Build the Future event was part of the pre-launch activities for the STS-133 mission. Image courtesy of Bill Ingalls/NASA.

Description of Teacher Participation and Activities: LEGO is also developing a website from which teachers can download all on-orbit video and corresponding activities. The information can be accessed at www.legospace.com. NASA and LEGO also co-sponsored several Build the Future events at Kennedy Space Center, where students, children, parents and teenagers are encouraged to build a model of a future rocket out of LEGOs.

Education Lead: Teresa Sindelar, NASA JSC, Houston, Texas, USA

Education Website: http://www.nasa.gov/education/tfs; http://www.legospace.com

Mission X: Train Like an Astronaut

Expeditions: Ongoing

Leading Space Agency: NASA, Global

Curriculum Grade Levels: K–8 (elementary), 9–12 (secondary)

Impact: This project involved 12 countries, 11 space agencies, 4164 students and 1 mission.

Participating Countries: Austria, Belgium, Colombia, Czech Republic, France, Germany, Italy, Japan, Netherlands, Northern Ireland, Russia, Spain, United Kingdom, United States

Number of K–8 Students (elementary), 9–12 Students (secondary): 4164

Description of Student Participation and Activities: Mission X is an international educational challenge that focuses on fitness and nutrition, teaching students how to train like an astronaut. Teams of students from the United States, Netherlands, Italy, France, Germany, Austria, Colombia, Spain and the United Kingdom are participating in this pilot project, while Japan hosts a modified version of the challenge. Russia, Belgium, the Czech Republic and Northern Ireland are supporting the implementation effort as observing countries. This first-of-its-kind global outreach project, sponsored by NASA's Human Research Program, brings together 11 space agencies and partner institutions to teach students principles of healthy eating and exercise. Mission X features training modules, such as FitKid, Astro Charlie and Walk-to-the-Moon, which compare to those used by astronauts and cosmonauts during all phases of space flight.

Education Lead: Chuck Lloyd, NASA JSC, Houston, Texas, USA

Education Website: http://trainlikeanastronaut.org

Students from Sharon, Massachusetts, participate in Mission X: Train Like An Astronaut. Photos courtesy of Ana Gonzalez.

Students from Sharon, Massachusetts, participate in Mission X: Train Like An Astronaut. Photos courtesy of Ana Gonzalez.

Table 17 – Countries and Number of Students Participating in Mission X: Train Like an Astronaut

Country	Lead Space Agency	Students	Teams	Partners
Colombia	Columbia Commission of Space (CCE)	810	40	Agustín Codazzi Geographic Institute (IGAC), Fundacion Cuidad Horizon 2050
USA	NASA	807	7	College Station Independent School District
United Kingdom	UK Space Agency	500	8	Venture Thinking, Royal Observatory Greenwich
Netherlands	ESA, Netherlands Space Office	490	21	
Spain	CDTI	359	18	Universidad Politecnica Madrid
Italy	ASI	300	7	Turin Planetarium (Infinito)
Germany	DLR	297	12	
Austria	Austrian Research Promotion Agency (FFG)	250	10	Planetarium Wien
France	Centre National d'Études Spatiales (CNES)	221	10	
Czech Republic	ESA	75	3	Czech Space Office
Japan	JAXA	598	4	Tsukuba Young Astronauts Club, Kanazawa Juichiya Elementary School, Waseda University and Miraikan
Belgium	ESA	25	1	
12*	**11**	**4164**	**137**	**9**

Russia and Northern Ireland were included in the working group as observing partners; they supported the effort as it developed, but did not host teams for the pilot.

Space Devices and Modern Technology for Personal Communication (MAI-75)

Expeditions: 11–28, Ongoing
Leading Space Agency: Roscosmos

Curriculum Grade Levels: K–8 (elementary), 9–12 (secondary), College (undergraduate)
Impact: This experiment attracted a large number of educational institutions and amateur operators in Russia and abroad. Sessions with the space station were conducted on a regular basis.

Participating Country: Russia

Description of Student Participation and Activities: The educational objectives of MAI-75 are to (1) check the personal communication between users and information resources on the ground and crews inside the ISS and (2) verify the ability to use standard Internet protocols to access information resources within the ISS. These objectives are included in the educational process related to higher profile aerospace information and telecommunication resources used in the experiment.

The use of images of Earth taken from space in the learning process is a promising application. In the experiment MAI-75, quick video from space in is downlinked and distributed in near-real time to face the challenges of educational institutions at various levels.

Education Lead: Sergey Avdeev, Roscosmos
Website: http://knts.tsniimash.ru/en

MAI Control Center for receiving and processing information seen during communications with the ISS. Image courtesy of Roscosmos.

Image taken by the reception and processing Center MAI. Image courtesy of Roscosmos.

Ten-Mayak (Shadow-Beacon)

Expeditions: 10–28, Ongoing

Leading Space Agency: Roscosmos

Curriculum Grade Levels: K–8 (elementary), 9–12 (secondary), College (undergraduate)

Participating Country: Russia

Description of Investigation: The goal of the experiment is to study very high frequency (VHF) radio reception and transmittal signal conditions from an onboard radio beacon located on the ISS Russian Segment (RS) using the world radio amateur network and determine the characteristics of the radio signals broadcasting and re-transmitting using an onboard transceiver. A secondary goal is to have reception of the input data and experience of operative work to use in the development of proposals concerning use of the space experiment Ten-Mayak technique to benefit educational programs.

The experiment tasks include (1) generating sounding impulses with an onboard radio beacon in the form of control time tags and signal reception by a ground reception network that includes amateur VHF receivers with subsequent transfer of measurement results via the Internet; (2) checking the principle of enforced blackout during short package generation under radio communication protocol AX25; (3) assessing the influence of sensitivity differences within the ground reception network on the accuracy of spatiotemporal measurements with the application of the given "board-ground" inter-space radio sounding technique; (4) checking technical capabilities and preconditions to use in the development of proposals for the creation of laboratory works based on the methodology of space experiment Ten–Mayak.

Description of Student Participation and Activities: The students engaged in this project study conditions of admission and transfer of the VHF-radio beacon onboard mode on the Russian

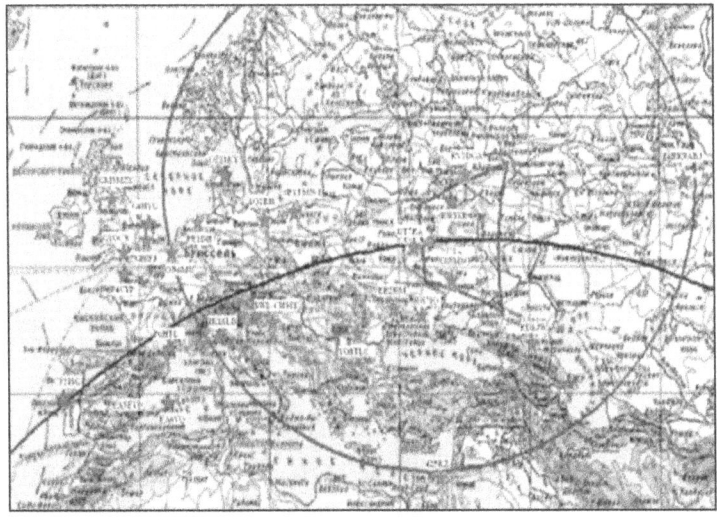

Locations of ground receiving stations.
Image courtesy of Roscosmos.

Segment using the world amateur radio network to define the characteristics and spatial distribution of the intensity of radio broadcast and rebroadcast using the onboard transceiver transmitter. Students also baseline data and operational experience to develop proposals on how to use the methodology of Ten-Mayak (Shadow-Beacon) for comprehensive programs, in particular those in the laboratory format. The methodology for Ten-Mayak (Shadow-Beacon) radio amateur packet-oriented communication provides opportunities for students to use the transport digital environment of the Internet to alert participants to experiments and to collect scientific information. The project Ten-Mayak (Shadow-Beacon) has received significant support from radio amateurs (with more than 200 applications for participation), which formed a continental dimensional field.

Principal Investigator: O. M. Alifanov, corresponding member RAS, Moscow Aviation Institute (MAI), Russia

Principal Investigator Website: http://knts.tsniimash.ru/shadow/en

The initiative of the Center for Social Assistance for Families and Children "Pechatniki" consists of a proposal for a new Ten-Mayak (Shadow-Beacon) space experiment application to be used for practical training in educational institutions. The implementation of collective sessions of proving signal reception, which can be particularly successful and effective, notably applies to extracurricular activities in secondary school.
Image courtesy of Roscosmos.

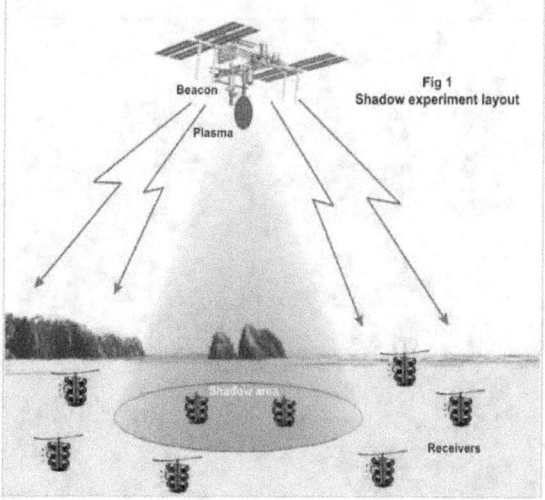

Diagram of the experiment Ten-Mayak (Shadow-Beacon).
Image courtesy of Roscosmos.

Try Zero-Gravity (Try Zero-G)

Expeditions: 19/20, 21/22, 23/24, 29/30, 31/32

Leading Space Agency: JAXA

Curriculum Grade Levels: K–8 (elementary), 9–12 (secondary)

Participating Country: Japan

Investigation Description: Try Zero-Gravity (Try Zero-G) allows the public, especially kids, to vote for and suggest physical tasks for JAXA astronauts to perform to demonstrate the difference between 0 g and 1 g for educational purposes. Some of tasks included putting in eye drops, performing push-ups on the ceiling, arm wrestling and flying a magic carpet.

Student Activity: Try Zero-Gravity (Try Zero-G) allows children to interact with ISS crewmembers through various activities for educational purposes. These activities help to enlighten the general public about microgravity and human space flight and demonstrate that microgravity is useful not only for scientists and engineers, but also for writers, poets, teachers and artists. The Try Zero-G activities are downlinked, edited and used to support educational resources for educators throughout Japan.

Principal Investigator: Naoko Matsuo, Japan Aerospace Exploration Agency, Tsukuba, Japan

HD video screen shot of Astronaut Koichi Wakata performing the Magic Carpet activity as part of the Try Zero-G experiment. Image courtesy of JAXA.

HD video screen shot of Astronaut Koichi Wakata performing the Water Pistol activity as part of the Try Zero-G experiment. Image courtesy of JAXA.

Cultural Activities

These educational activities provide a cultural experience for students and link science with the humanities.

Great Start

Expeditions: 27/28, 29/30, Ongoing

Leading Space Agency: Roscosmos

Curriculum Grade Levels: K–8 (elementary), 9–12 (secondary), College (undergraduate)

Participating Country: Russia

Description of Student Participation and Activities: This is an experiment aimed at popularizing the achievements of cosmonautics in Russia and in the world. A special questionnaire was developed to allow students, educators and the general public to express their attitudes about a great event in human history — the first human flight in space — as well as to get acquainted with the results of experiments and to participate virtually in today's space program.

This experiment promotes and enhances international cooperation on the ISS to further integrate Russia into the world of cultural, educational and scientific relations. As a result, the experiment holds scientific-educational workshops and events to popularize the achievements of Russian manned space and, consequently, brings the exploration of space to the general population — students, teachers, specialists and researchers in various areas of possible utilization of the results of space missions.

Education Lead: M. U. Belyaev, S. P. Korolev RSC "Energia," Russia

Education Website: http://gagarin.energia.ru/

Russian students. Image courtesy of Xinhua.

Members of Expedition 29/30 to the International Space Station, U.S. astronaut Dan Burbank, left, and Russian cosmonauts Anton Shkaplerov, right, and Anatoly Ivanishin gesture after a news conference at the Baikonur Cosmodrome in Kazakhstan, Saturday, November 12. The crew blasted off for the International Space Station Monday, November 14, 2011, aboard a Russian-made Soyuz spacecraft, demonstrating international cooperation — one of the benefits of the space station. Image courtesy of Associated Press.

JAXA Uchu Renshi (Space Poem Chain)

Expeditions: Ongoing

Leading Space Agency: JAXA, Global

Curriculum Grade Levels: K–8 (elementary), 9–12 (secondary)

Participating Country: Japan

Number of K–8 Students (elementary): 560
Number of 9–12 Students (secondary): 60
Number of Schools: 15
Number of Teachers: 20

Description of Student Participation and Activities: From the beginning of recorded time, the skies above have been sometimes a field on which human beings painted their dreams, sometimes a mirror for our own lives. Today, when scientific progress continues to unravel the world's mysteries one by one, the skies above, extending to outer space, continue to inspire in us a limitless curiosity and at the same time school us in awe for the infinite. The Space Poem Chain is an attempt to create a collaborative place through "linked verse" by thinking together about the universe, Earth, and life itself, unfettered by barriers of nation, culture, generation, profession, and position or rank.

Chain poetry itself (renshi), a form developed from traditional Japanese linked verse (renga and renku) by the members of the Kai group, including poet and critic Makoto Ooka and poet Shuntaro Tanikawa, is known and practiced almost worldwide. Space Poem Chain Volume 3 was compiled, with the supervision of Ooka and the contributions of Tanikawa, from entries contributed over the Internet by the general public combined with contributions from poets and other cultural figures.

Astronaut Takao Doi on board the station with a DVD containing the Space Poem. NASA image S123E007866

During the first phase (October 2006 to March 2007) we combined public submissions and contributions to create a sequence 24 poems long. The first poem was contributed by the JAXA astronaut Naoko Yamazaki, and after that we received submissions from approximately 800 people from Japan and abroad; contributors ranged in age from 8 to 98. Planetariums and schools also began to incorporate the Space Poem Chain into their programs. The completed poem chain was read at the Space Poem Chain Symposium and then recorded on a DVD, which was entrusted to astronaut Takao Doi in March 2008 and stored in the International Space Station's Japanese Experiment Module, Kibo (Hope). While the space station was in orbit, astronaut Doi took a commemorative photograph of the DVD.

During the second phase (July 6, 2007 to February 8, 2008), we compiled another Space Poem Chain of 24 poems titled "There Are Stars." Among the contributors were poets from Asia and the Pacific region, and submissions were sent by people in many countries around the world. The number of planetariums and schools in Japan and abroad that integrated the Space Poem Chain into their curricula also increased. Now that the complete 24-poem chain has been made public at the Space Poem Chain symposium, it will be uploaded and stored on Kibo by astronaut Koichi Wakata.

During the third phase (September 5, 2008 to April 10, 2009), we again looked forward to a 24-poem Space Poem Chain. By expressing thoughts and feelings about the universe and life in poetry, we hope to create a space in which imagination can flourish. Fortunately, linked verse, growing as it does out of a collaborative interweaving of minds, has an organic quality that lends itself well to these aims. To help the process along, we added a comment corner that will explain the choices of poems and add thoughts about the next poem to come. We invited young poets and Koichi Wakata, a JAXA astronaut aboard Kibo during the long-term expeditions, to contribute. This was the first mission in which the Space Poem Chain was compiled with astronauts aboard the station, and the mission was successfully accomplished. Many people enjoyed Wakata's video images from Kibo.

For the fourth phase (fiscal year 2010), Ryoko Shindo, a poet, was invited to participate as a supervisor, and she compiled a Space Poem Chain incorporating verses written by children and adults. We would like to explore the universe in children or a children's universe, led by them. The completed Space Poem Chain was launched in 2011.

Education Leads: Teruhiko Tabuchi

Education Websites: http://iss.jaxa.jp/utiliz/renshi/index.html

Summary of International Space Station Education Opportunities

Table 18 – Student, Teacher and School Summary of ISS Education Opportunities

Investigations	Number of Students				Schools	Teachers
	K-12	Undergraduate	Graduate	Postdoctoral		
Student-Developed Investigations	557,312	326	9	0	17,273	19,308
Education Competitions	3806				74	77
Students Performing Classroom Versions of ISS Investigations	119,425	8	8		1989	2000
Students Participating in ISS Investigators Experiments	31,607	1364	1099	26	35	5017
Students Participating in ISS Engineering Education: Hardware Development	2116	273	46		6	63
Educational Demonstrations and Activities	41,464,680				2,637	2,796,820
Cultural Activities	620				15	20
Totals	~42,180,000	~2,000	~1,100	~26	~22,000	~2,800,000

International Space Station Education Accomplishments

These educational activities and projects were designed and conducted from the first element launch of the International Space Station to Expedition 27/28, but are no longer actively involving students.

Student-Developed Investigations

These experiments typically were developed by students under the aegis of a teacher or scientist mentor and performed solely for the benefit of the students.

Analysis of Inertial Solid Properties (APIS)

Expeditions: 7, 8
Leading Space Agency: ESA
Grade Levels: 10–12

Investigation Description: The main scientific objective of the experiment was to demonstrate solid body rotation principles and to prepare a video for later use in education. The experiment was focused on the behavior of a rigid body rotating around its centre of mass. The experimental configuration was seen as an approximation of a much more complex system, a rotating spacecraft (which can be considered, from the mechanical point of view, to be composed of a number of rigid bodies joined by complex links). To keep efforts required for the analysis reasonable, experiments were concerned only with torque-free rotational motion. The experiments intended to show the different types of motion that appeared with changes in the mass distribution of the body (that is, changes that depended on the inertia tensor structure: spherical, cylindrical or ellipsoidal) and changes in the mechanical energy dissipation effect caused by external or internal actions. An undesirable effect of internal energy dissipation was a change in the axis of rotation, such as that experienced by the first American satellite, Explorer I. Similar changes to the spin axis also appeared during mass distribution changes (e.g., deployment of booms, solar panels, and antennae).

Principal Investigator: Ana Laveron Simavilla, Universidad Politecnica de Madrid, Madrid, Spain

Biological Research in Canisters – 16: Actin Regulation of Arabidopsis Root Growth and Orientation During Space Flight (BRIC-16-Regulation)

Expeditions: 23/24
Leading Space Agency: NASA
Grade Level: College (undergraduate)

Investigation Description: This investigation studied how actin cytoskeleton dictates root growth orientation during space flight and involved an extensive set of genome-wide microarray studies to unravel actin-dependent gene regulatory networks that modulate root growth and orientation during space flight.

Principal Investigator: Elison Blancaflor, Ph.D., Samuel Roberts Noble Foundation Incorporated, Ardmore, Oklahoma, USA

Drop Your Thesis! 2011: Falling Roots

Leading Space Agency: ESA

Grade Level: College (undergraduates)

Investigation Description: The Falling Roots team comprises two students from the University of Florence, Italy, who have been selected to develop and perform a microgravity experiment during ESA's Drop your Thesis! 2011 campaign. Their experiment is a continuation of their investigation into how biological processes are affected by microgravity and hypergravity. Plants have evolved under the constant force of gravity, and its presence strongly influences their growth and development. For this reason, a change in gravitational field strength can be considered a source of stress that is perceived in the plant's roots and then transmitted to the other organs by signaling chains, resulting in an adaptation of plant physiology.

Student Activity: In this experiment, several roots were placed, together with water, in an array of syringes. The experimental setup was placed in a capsule and suspended at the top of the ZARM drop tower. A few minutes prior to the drop, a plunger depressed several syringes, and control samples were collected in small containers. These control samples were later used to determine the level of reactive oxygen species production in roots that have been subjected to all of the same conditions as the test samples, except for microgravity. Comparison of the concentration of compounds in the test samples with that in the control samples permitted the acquisition of scientifically significant results on the production of signaling molecules in microgravity conditions.

Principal Investigator: Stefano Mancuso, University of Florence, Florence, Italy

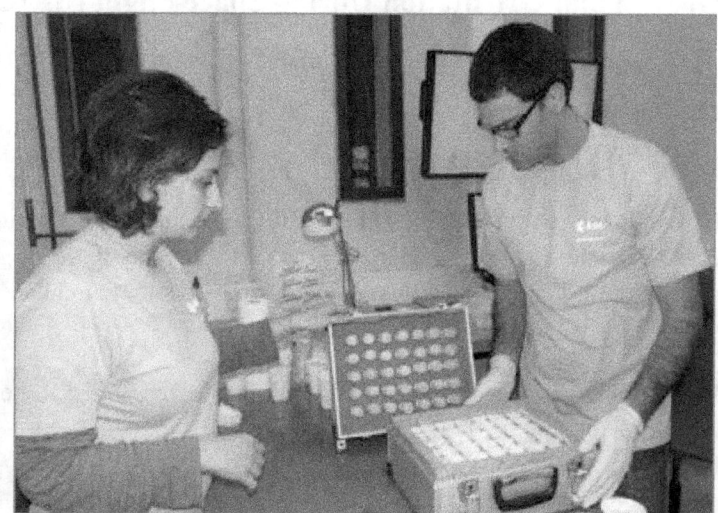

The Falling Roots team.
Image courtesy of N. Callens/ESA.

Drop Your Thesis! 2010: Bubble Jet Impingement in Microgravity Conditions

Leading Space Agency: ESA

Grade Level: College (undergraduates)

Investigation Description: A group of three students from Universitat Politècnica de Catalunya (UPC) developed an experimental setup to investigate the collision of bubble jets in microgravity. The experiment was carried out at the Center of Applied Space Technology and Microgravity (ZARM) drop tower in Bremen, Germany. The results could improve our understanding of how two-phase flows behave in a low-gravity environment. In particular, the effects of gas-liquid flow rates and separation between jets on the bubble coalescence probability will be investigated.

Student Activity: In this experiment, two opposed bubbly jets were introduced in a tank initially filled with a quiescent liquid. The data obtained from this experiment will complement the work of Carrera et al., performed at the ZARM drop tower in 2003, in which the structure of a single bubbly jet was investigated. The results of this experiment will help to improve our general understanding of two-phase flows, in particular the bubble generation process, bubble-bubble interactions (coalescences, bouncing), and the jet-collision phenomenon in a low-gravity environment.

Principal Investigator: Ricard González Cinca, Applied Physics Department, UPC, Barcelona, Spain

The Bubble Jet team. Image courtesy of ESA.

Drop Your Thesis! 2009: New Polymer Dispersed Liquid Crystal (PDLC) Materials Obtained From Dispersion of Liquid Crystal in Microgravity Conditions

Leading Space Agency: ESA

Grade Level: College (undergraduates)

Investigation Description: Porto (FCUP) used the ZARM drop tower in Bremen, Germany, to conduct an experiment involving polymerization of different types of PDLCs in microgravity conditions. The polymerization was achieved using ultraviolet illumination. PDLCs consist of micron-sized droplets of liquid crystal dispersed in a polymer matrix. This mixture combines the mechanical properties of polymers with the electromagnetic properties of liquid crystals, allowing control of the optical anisotropy of the PDLC film through the application of a specific electrical or magnetic field.

Student Activity: In this experiment, three different mesophases of liquid crystal were used: the nematic E7 from Merck, the chiral nematic E7+CB15 from Merck, and the chiral smetic C Felix 017/100 from AZ-Electronics. Each was mixed with a photopolymer, NOA-81 (Norland). The resulting mix was inserted into glass cells, where it was pre-cured in microgravity. The remaining full-cure process was performed in normal gravity conditions. In addition, some samples were polymerized with induced inhomogeneities. These were expected to provide unique structures that can be produced only in a microgravity environment.

Principal Investigator: Carla Carmelo Rosa, Physics Department, Faculdade de Ciências da Universidade do Porto (FCUP), Porto, Portugal

Educational Demonstration of Basic Physics Laws of Motion (Fizika-Obrazovanie)

Expeditions: 18, 19/20, Ongoing
Leading Space Agency: Roscosmos
Grade Levels: 7–12

Investigation Description: Integrated experiment FIZIKA-OBRAZOVANIE consisted of three sections: FIZIKA-LT, FIZIKA-FAZA and FIZIKA-OTOLIT. In the experiment, the following physical phenomena were proposed to be demonstrated in microgravity: reactive and aerodynamic action on a solid body of revolution (a series of experiment sessions using a Letauschaya Tarelka device), aggregation of gas bubbles during the phase separation of a gas-liquid fine-dispersion medium (an experiment session using a Faza device), and processes of the motion and effect transfer to the human vestibular apparatus (a series of experiment sessions using an Otolit device).

Principal Investigator: N.L. Shoshunov, S. P. Korolev, RSC "Energia," Russia
Website: http://knts.tsniimash.ru/en

Fizika-Obrazovanie space experiment equipment.

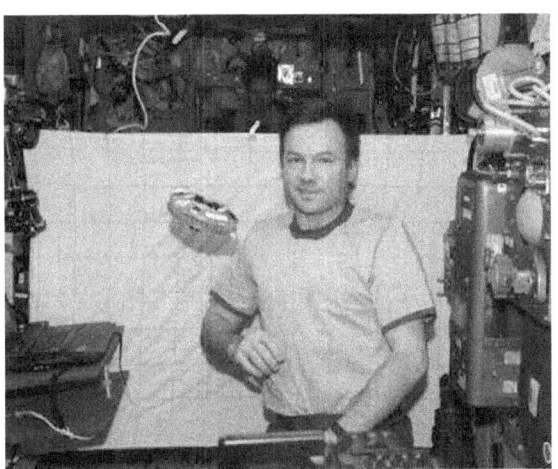

Russian cosmonaut Yu. V. Lonchakov (onboard engineer on Expedition 18) carries out the Fizika-Obrazovanie SE television session using Letauschaya Tarelka equipment.

Russian cosmonaut G. I. Padalka (commander of Expedition 19) performs an experiment using Letauschaya Tarelka equipment.

Students of Moscow lycée No. 1537 during the Earth–ISS onboard television session in the Russian Mission Control Center (MCC-M).

Electrostatic Self-Assembly Demonstration (ESD)

Expeditions: 10, 11
Leading Space Agency: ESA
Grade Levels: 7–12

Investigation Description: This educational activity involved the self-assembly of components larger than molecules into ordered arrays, an efficient way of preparing microstructured materials with interesting properties. Electrostatic self-assembly occurs when different types of components charge with opposite electrical polarities. The interplay of repulsive interactions between like-charged objects and attractive interactions between unlike-charged objects results in the self-assembly of these objects into highly ordered, closed arrays. This educational experiment demonstrated the electrostatic self-assembly of two different types of macroscopic components or spheres of identical dimensions to create different molecular structures in weightless conditions on board the station; three demonstrations were filmed. For all in-space demonstrations, comparable on-ground experiments were performed and filmed to familiarize students with the differences between the Earth and space environments.

Student Activity: Videos of both the on-ground and in-space demonstrations were recorded, and the footage was used to develop a ISS DVD lesson fitting the basic European science and technology curriculum of the target 12- to18-year-old age group. The DVD was distributed in 12 languages to secondary school teachers in ESA member states.

Principal Investigator: William Carey, Erasmus User Centre and Communication Office, Noordwijk, Netherlands

Fly Your Thesis! 2009: ABC Transporters in Microgravity

Leading Space Agency: ESA
Grade Level: College (undergraduates)

Investigation Description: A team comprising two students from the Autonomous University of Barcelona and two from the Polytechnic University of Catalonia developed an experiment to investigate the behavior of certain biological agents involved in the assimilation of drugs by the human body. The results could help to improve medical treatments in space. ATP-binding cassette (ABC) transporters are enzymes present in all cells of the human body. Their main function is to remove toxic waste and drugs from the body. Unfortunately, this means that they have a negative impact on current drug treatments, especially in cancer cells, where they are a major cause of resistance to tumor therapy.

Student Activity: The aim of this project was to develop technology that could make it possible to study drug behavior in space without testing humans directly. It involved study of the activity of an ABC transporter model known as Multidrug Resistance-Associated Protein 2 (MRP2) in microgravity conditions. The experiment worked well during the parabolic flight campaign and provided the student team with a large number of samples to be analyzed. The performance of MRP2 was then compared with similar control experiments on Earth to ascertain whether MRP2's function was altered by microgravity.

The ABC team working on its experiment.
Image courtesy of ESA.

The results of this research may lead to a better understanding of the ABC transporters and their role in space and terrestrial pharmacology. By providing new information and introducing new research methods involving microgravity, the project may eventually help to improve treatments for cancer patients and foster research into drugs that are suitable for use on future moon and Mars missions.

Principal Investigator: Felip Fenollosa Artés, Polytechnic University of Catalonia and Autonomous University of Barcelona

Fly Your Thesis! 2009: AstEx – Simulating Asteroidal Regoliths: Implications for Geology and Sample Return

Leading Space Agency: ESA

Curriculum Level: College (undergraduates)

Investigation Description: The AstEx team comprised three students — two from the Open University in the United Kingdom and one affiliated with both the Open University and the University of Nice-Sophia Antipolis in France. The aim of the experiment was to study the behavior of granular material in microgravity; possible applications include the design of future asteroid sample return missions and the interpretation of asteroid geology. There is still some uncertainty about how such asteroidal regolith material behaves in this low-gravity regime. The AstEx experiment investigated how a steady-state (constant) flow could be achieved in a granular material in microgravity conditions.

Student Activity: Glass beads were tracked during the period of microgravity available during a single parabola flown by the Airbus A300 aircraft. By calculating the velocities of the glass beads, the time required to initiate a steady-state flow was determined. Comparisons with ground-based experiments should show how a steady-state flow in microgravity differs from a steady-state flow on Earth. Another investigation was carried out to determine what effect reversing the direction of shear had on the steady-state flow, and the results are being compared with ground-based results. Each of the studies was repeated with glass beads of different sizes and with different shear rates. The AstEx experiment worked well during the parabolic flight campaign and provided the student team with a very large amount of data to be analyzed.

Principal Investigator: Dr. Simon F. Green, The Open University, United Kingdom, and the University of Nice-Sophia Antipolis, France

Fly Your Thesis! 2009: Complex — Microgravity Studies of the Effect of Volume Fraction and Salinity on Flow in Samples of Clay Nanoparticles Dispersed in Water

Leading Space Agency: ESA
Grade Level: College (undergraduate)

Investigation Description: Complex comprised three students from the University of Science and Technology in Trondheim, Norway. The aim of their experiment was to study the flow of clay nanoparticles in salty water to provide a deeper understanding of the self-organization of such small particles. This experiment concerned the basic properties of clay nanoparticles. It investigated flow-induced order in aqueous suspensions of synthetic smectite clay, sodium fluorohectorite. In particular, it studied the decay of this order in microgravity conditions due to Brownian rotational motion. Performing the study in a weightless environment was important because it eliminated problems related to sedimentation and convection that could interfere with the pure Brownian dynamics of the system.

Student Activity: During the parabolic flight campaign, the Complex experiment was successful and provided the student team with interesting data to be analyzed. The team's research may have several practical applications. For example, a more fundamental understanding of the properties of nanoparticles in flow is essential in key industries, such as storage of nuclear waste, crude oil extraction, and the development of new, smart nanoapplications. In addition, the role of flow and orientation of clay particles in soil stability is not yet properly understood, so this experiment may help to prevent catastrophic landslides.

Principal Investigator: Prof. Dr.-Ing. Jon Otto Fossum, Norwegian University of Science and Technology, Trondheim, Norway

Fly Your Thesis! 2009: Dust Side of the Force — The Importance of the Temperature Gradient Effect in Planet Formation Processes and Dust Devils or Storms on Mars

Leading Space Agency: ESA

Grade Level: College (Undergraduate)

Investigation Description: The Dust Side of the Force was a team of five students from the Institute of Planetology at the University of Münster, Germany. The objective of their experiment was to investigate the force induced by a temperature gradient in dust beds, which is thought to be important in planet formation and the formation of dust storms on Mars.

Student Activity: The experiment comprised an evacuated vacuum chamber in which a sample of simulated Martian soil was placed and then illuminated by a high-power halogen lamp. The induced dust particle ejections were observed and recorded in microgravity, 1 g and 2 g conditions. The gravitational dependency of the ejection forces was then reconstructed. The experiment worked successfully during all the parabola phases and provided the student team with a large number of videos to be analyzed. The research is relevant to understanding the deposition of dust into the Martian atmosphere and in protoplanetary disks, such as the solar nebula from which the Earth and other planets formed about 4.5 billion years ago.

Principal Investigator: Prof. Dr. Gerhard Wurm, Faculty of Physics, University of Duisburg-Essen, Essen, Germany

The Dust Side of the Force team working on its experiment.
Image courtesy of ESA.

Foam Optics and Mechanics – Stability (Foam-Stability)

Expeditions: 19/20, 21/22
Leading Space Agency: ESA
Grade Levels: 10–12

Investigation Description: Foam-Stability was an investigation of aqueous and non-aqueous foams in microgravity. The behavior of foams in microgravity is very different from their behavior on Earth because the process of drainage is absent in microgravity.

Student Activity: Students explored the following fundamental questions: How long can those foams be stable? What is the role of solid particles in the liquid in water foam stabilization? Is it possible to create very "wet" foams in microgravity?

Principal Investigator: Pr. Dominique Langevin, Universite Paris-Sud, Orsay, France

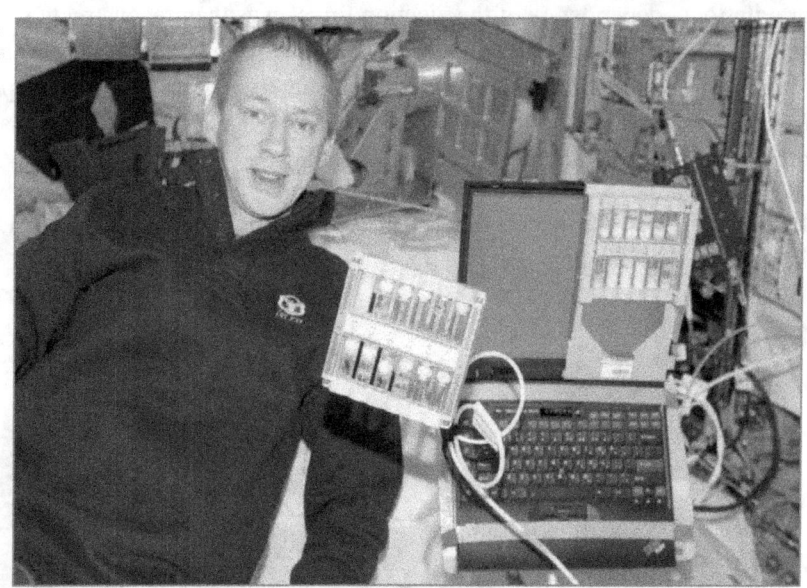

Expedition 20 flight engineer Frank De Winne is pictured near an ESA Foam-Stability experiment floating freely in the Columbus laboratory of the International Space Station. Image ISS020E042309.

Image Reversal in Space (Iris)

Expeditions: 19/20
Leading Space Agency: CSA
Grade Levels: 7–12

Investigation Description: The Iris experiment was an educational experiment developed by students at the International Space University. This experiment studied the effects of microgravity on the way people perceive two-dimensional (2D) and 3D objects. The experiment and its software were designed by a multidisciplinary group of students, giving them valuable experience as they prepared to launch their own careers in space.

Student Activity: Iris was developed in the summer of 2008 by several students at the International Space University when they were charged to create a simple experiment that could be performed in space. The purpose was to teach the students how to design an experiment for testing in microgravity.

Principal Investigator: Gilles Clement, Ph.D., Centre National de la Recherche Scientifique, Toulouse, France

Space Experiment Module (SEM)

Expeditions: 10, 11, 13, 14
Leading Space Agency: NASA

Grade Levels: 7–12

Impact: Eleven schools and 3300 students were engaged in running experiments on the first SEM satchel flight.

Investigation Description: The SEM introduced students to the concept of performing space-based research on the International Space Station. It provided students the opportunity to conduct their own research on the effects of microgravity, radiation and space flight on various materials.

Student Activity: Research objectives for each experiment were determined by the students, but generally included hypotheses on changes in the selected materials caused by the space environment.

Students were provided space capsules to contain passive test articles for flight. The capsules were then packed in satchels (10 per satchel) containing formed foam layers for flight.

Principal Investigator: Ruthan Lewis, Ph.D., NASA Goddard Space Flight Center, Greenbelt, Maryland, USA

Test of the Basic Principles of Mechanics in Space (THEBAS)

Expeditions: 7, 8
Leading Space Agency: ESA
Grade Levels: 7–12

Investigation Description: The objectives of the experiment were to illustrate with relatively simple hardware the principles of dynamics, ranging from the classical, rational mechanics of solid bodies to the continuous media mechanics, and to prepare a video for later use in education. The students analyzed the behavior of transparent, closed containers of the same size and total mass that were filled with solid bodies (spheres) of different radii. The mass of the contents of each container was the same in all considered cases. The experiment studied the interaction of a container and the particles inside it when the system is periodically oscillated in one dimension. In addition to the experiment performed in space, reference experiments with identical hardware were performed on the ground (being the displacement perpendicular to gravity) to quantify the effect of gravity on the system.

Principal Investigator: Ana Laveron Simavilla, Universidad Politecnica de Madrid, Madrid, Spain

University Research Centers–Microbial-1 (URC-Microbial-1)

Expeditions: 21/22
Leading Space Agency: NASA
Grade Level: College (undergraduate)

Investigation Description: University Research Centers–Microbial-1 (URC-Microbial-1) evaluated morphological and molecular changes in *Escheria coli* and *Bacillus subtilis* in microgravity. This investigation was a proof-of-concept model providing space flight experience to stimulate and excite University Research Center scientists and students at Texas Southern University. The experiment involved the study of morphological and molecular changes of *Escheria coli* and *Bacillus subtilis* microbes brought about by the space flight environment.

Each component of the experiment was designed to be reproduced easily in the classroom, providing hands-on experience to the students who were involved. Postflight repeat experiments were conducted in schools across the nation, and samples from this flight experiment were shared among principal investigators in other institutions. Texas Southern University developed courses incorporating the data into the microbiology curriculum.

Principal Investigator: Olufisayo Jejelowo, Ph.D., Texas Southern University, Houston, Texas, USA

Students in the Center for Bio-nanotechnology and Environmental Research at Texas Southern University prepare samples for the URC-Microbial-1 investigation. Image courtesy of Texas Southern University.

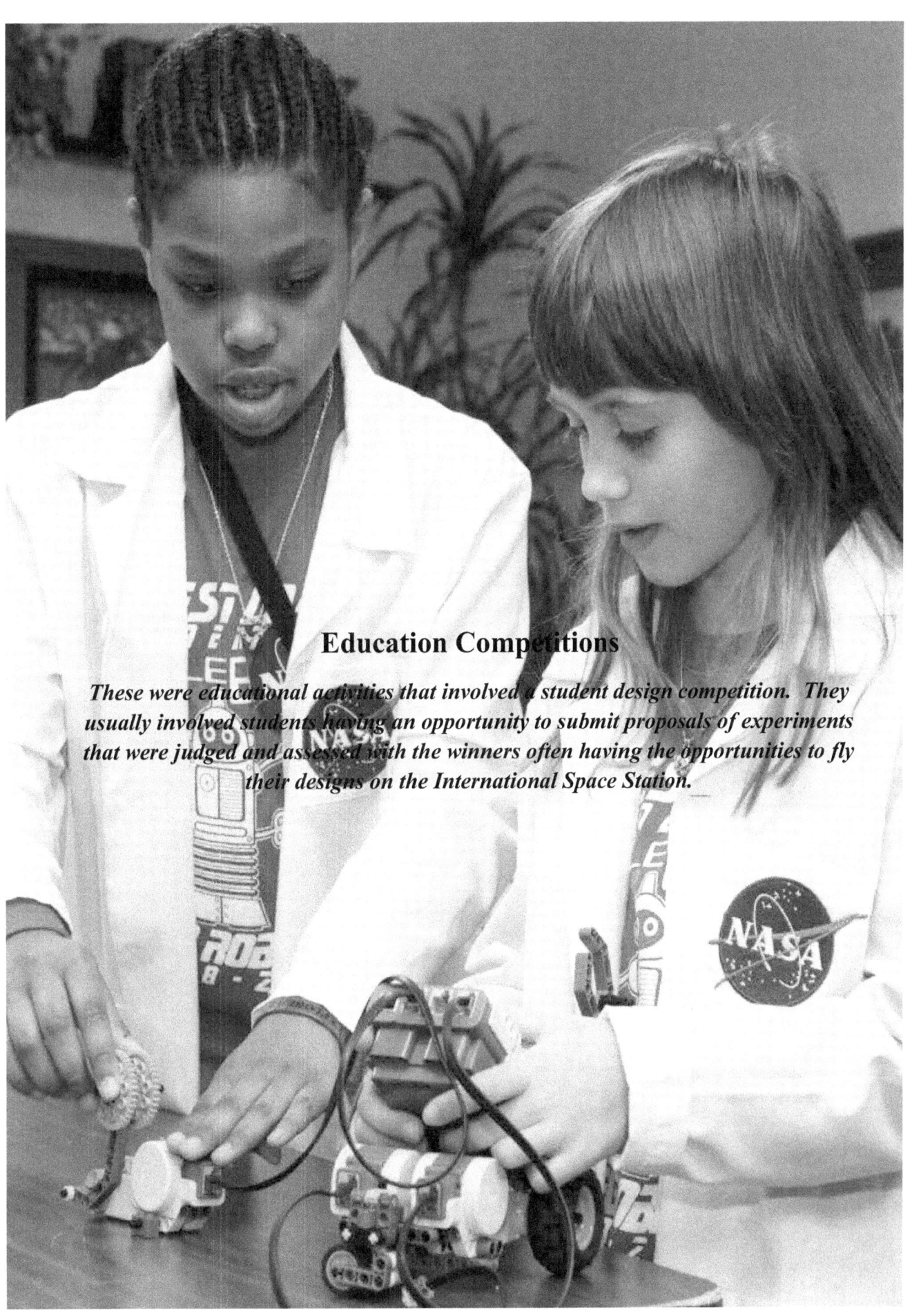

Education Competitions

These were educational activities that involved a student design competition. They usually involved students having an opportunity to submit proposals of experiments that were judged and assessed with the winners often having the opportunities to fly their designs on the International Space Station.

SUCCESS Programme

Expeditions: 2002–2010
Leading Space Agency: ESA

Impact: The investigation attracted hundreds of applications from students, and a dozen students were selected to have their payloads developed and their experiments operated in space or to have an internship at ESTEC, ESA's technological center.

Investigation Description: Space Station Utilization Contest Calls for European Student initiativeS (SUCCESS) was a competition that called for European university students of all disciplines to propose an experiment that could fly on board the International Space Station. The goal of the competition was to make today's students the International Space Station users of tomorrow. The first prize of the competition was a one-year internship at ESA's space research and technology centre, ESTEC, in the Netherlands or the opportunity to participate in designing the hardware for the proposed experiment.

Education Lead: Nigel Savage, Ph.D., ESTEC, Netherlands
Website: http://www.esa.int/esaHS/SEMU9TGHZTD_education_0.html

SUCCESS Bug Energy — Study of Output of Bacterial Fuel Cells in Weightlessness

Expeditions: 8, 9
Leading Space Agency: ESA
Grade Levels: K–8 (elementary), 9–12 (secondary)

Investigation Description: The BugNRG experiment studied the influence of weightlessness on the output of bacterial fuel cells. The aim was to acquire precision data on the output during a total drain of a fuel cell in weightlessness. In the right circumstances, bacteria are capable of transferring carbohydrates into free electrons, protons and carbon dioxide. This process was performed in bacterial fuel cells. When placed inside a two-chamber fuel cell, bacteria produced an electrical current, and the output was measured and recorded to enable study of the properties of the fuel cell. BugNRG was one of two experiments that won a student competition organized by the Dutch government through the Department of Education, Culture and Science.

Principal Investigator: Sebastiaan de Vet, Technical University of Delft, Delft, Netherlands

SUCCESS Chondro – Study on the Development of Methods To Produce Artificial Cartilage

Expeditions: 7, 8
Leading Space Agency: ESA
Grade Level: College (undergraduate)

Investigation Description: People all over the Earth are suffering from cartilage problems. Modern medicine is trying to develop methods to produce cartilage artificially so it can be implanted in humans. The influence of gravity, however, disturbs the process of cartilage structure formation in such a way that the structure does not completely satisfy the needs of today's medicine. The objective of the microgravity experiment was to find a more stable cartilage structure and to test the experiment hardware.

Student Activity: Students designed the incubator that contained the nutrient and cartilage cells.

Principal Investigators: Georg Keller, Vlada Stamenkovic, ETHZ University, Zurich, Switzerland

Cell Cultivation System flown on the station during Expedition 7. Image courtesy of the Space Biology Institute, ESA.

SUCCESS Graphobox – Study Into the Interaction of Effects of Light and Gravity on the Growth Processes of Plants

Expeditions: 8, 9
Leading Space Agency: ESA
Grade Level: College (Undergraduate)

Investigation Description: The GraPhoBox experiment investigates the presence of a link between phototropism (growth toward a light source) and gravitropism (growth toward the gravitational vector) in wild and mutant seeds of *Arabidopsis thaliana*.

Student Activity: Students designed and developed the experiment and performed analysis on the ground.

Principal Investigator: Karel Buizer, University of Utrecht, Utrecht, Netherlands.

SUCCESS UTBI – Under the Background Influence

Expedition: 14
Leading Space Agency: ESA
Grade Levels: College (undergraduate)

Investigation Description: The main goal of UTBI was the measurement of the background radiation inside spacecraft. Measurements of the X-ray, gamma ray and other particles (protons, neutrons, electrons) have a very important effect outside the Earth's geomagnetic field that can help in the development of future space vehicles and other space technologies.

Student Activity: University students created a new detector to measure radiation on board the International Space Station.

Understanding the effect of radiation on humans is an important factor that must be addressed before long-duration missions to the moon and Mars are planned.

Principal Investigator: Andres Russu, Universidad de Valencia, Valencia, Spain

SUCCESS Winograd – Effects of Gravity on Bacterial Development

Expeditions: 7, 8
Leading Space Agency: ESA
Grade Levels: College (undergraduate); Graduate (master's, Ph.D., M.D.)

Investigation Description: The Winograd experiment was used to grow Winogradski columns in a weightless environment. A Winogradski column is a colony of different types of bacteria wherein the waste products of one bacterium serve as the nutrients of the other. The systems were found in ordinary pond or lake water and needed no other input than light for photosynthesis.

Student Activity: The returned samples were analyzed by students to determine where certain bacteria were located during flight and hence determine the effect of weightlessness on the formation of Winogradski columns. The experiment clarified whether the bacteria in the water organized themselves in a pattern similar to the pattern they would form on Earth.

Principal Investigator: Darren Smillie, Tamara Banerjee, and Rishi Dhir, University of Edinburgh, Edinburgh, Scotland

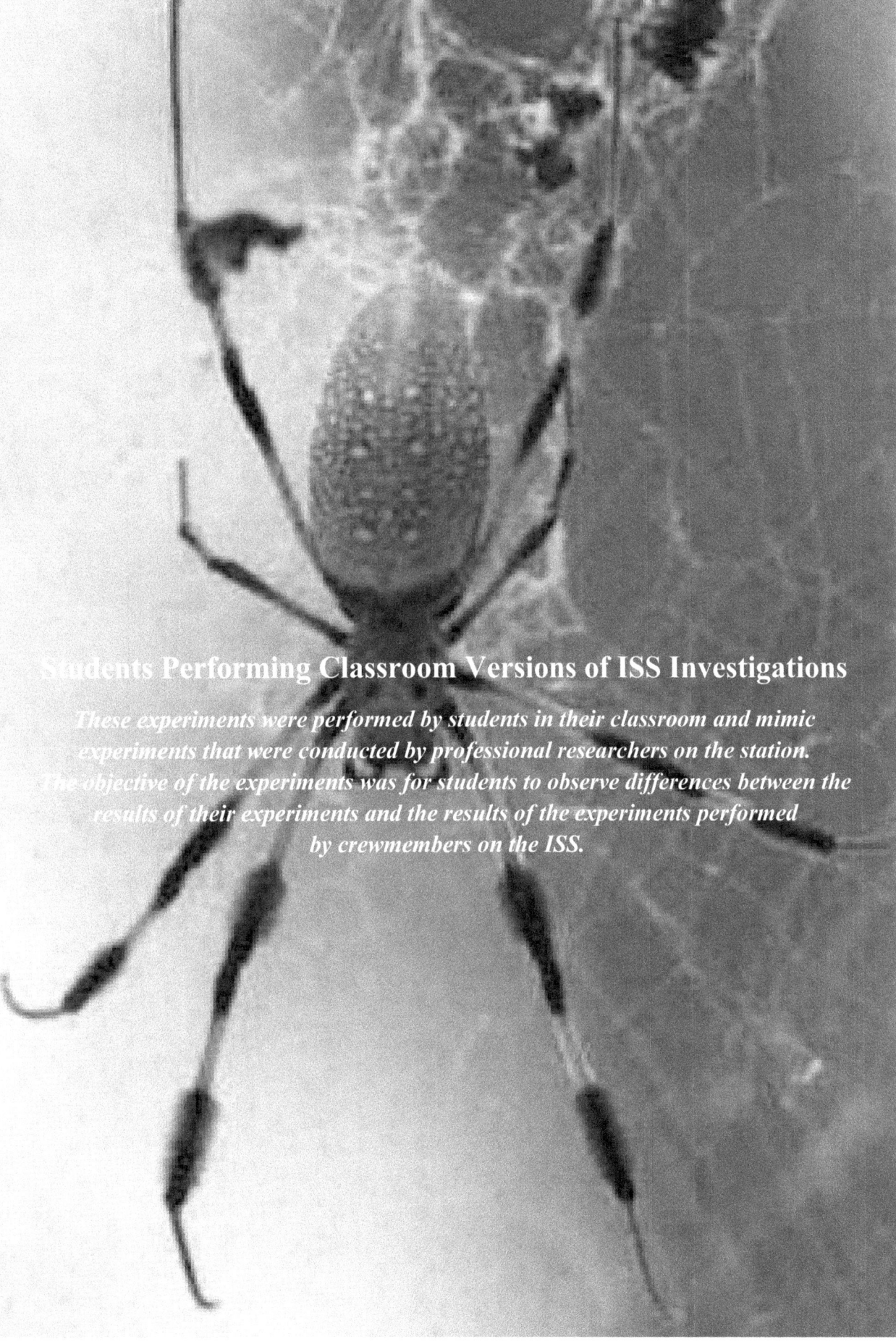

Students Performing Classroom Versions of ISS Investigations

These experiments were performed by students in their classroom and mimic experiments that were conducted by professional researchers on the station. The objective of the experiments was for students to observe differences between the results of their experiments and the results of the experiments performed by crewmembers on the ISS.

Advanced Astroculture (ADVASC) and Photosynthesis Experiment and System Testing and Operation (PESTO)

Expeditions: 2, 4, 5
Leading Space Agency: NASA
Grade Levels: K–8 (elementary), 9–12 (secondary)

Investigation Description: Advanced Astroculture™ (ADVASC) was a commercially sponsored payload that provided precise control of environmental parameters for plant growth, including temperature, relative humidity, light, fluid nutrients and atmosphere. ADVASC hardware was used in a series of tests performed over the course of three different expeditions. On Expedition 2, ADVASC demonstrated the first "seed-to-seed" experiment in space by growing *Arabidopsis thaliana* (thale cress) through a complete life cycle. On Expedition 4, 35 percent of the space-grown seeds and 65 percent of the wild *Arabidopsis* seeds were grown on orbit, with samples preserved for genetic analysis. Finally, on Expedition 5, soybean plants were grown through an entire life cycle, from seed to seed.

Student Activity: The Orbital Laboratory was an Internet-based multimedia tool that allowed students and educators to conduct their own ground-based plant experiments and to analyze data returned from the ADVASC units in their classrooms on Earth. The student activities for Expedition 2 focused on plant growth. As the seeds grew in ADVASC on ISS, students on Earth were growing seeds in their own plant growth systems. The on-orbit Expedition 4 activities used seeds harvested from the first generation of plants that grew on ISS. Focusing on genetics, the students grew the same seeds on Earth along with four mutations of the original seed.

Principal Investigator(s): Weijia Zhou, University of Wisconsin – Madison, Madison, Wisconsin; Tom Corbin, Pioneer Hi-Bred International, Inc.(ADVASC); and Gary Stutte, Dynamac Corporation, NASA KSC, Cape Canaveral, Florida (PESTO), USA

Agrospace Experiment Suite (AES)

Expedition: 10
Leading Space Agency: ESA
Grade Levels: 5–12

Investigation Description: The aim of this part of the experiment, Space Beans for Students, was to involve and interest children in space science to help with the continuous exploitation of space technology and its application to everyday life on Earth. To this end, the experiment consisted of beans germinating in space at the same time that beans were germinating in classrooms on Earth. The germination of seeds in space had been demonstrated previously, but this experiment involved students in the mission and increased their knowledge of the space environment and the potential applications of space technology.

Student Activity: Most students are already aware that plants begin their life cycle when a dormant seed germinates and a seedling begins growing. In this activity, students investigated the conditions that lead to the germination of seeds and observed in parallel the early stages of the plant life cycle on Earth and in space. The main purpose of the experiment was to provide the student with answers to the following key questions: What do plant seeds need to germinate? What changes do seeds and seedlings undergo during germination and early growth? What are the main differences between seeds germinating on Earth and those germinating in the weightlessness of space?

Principal Investigator: Gui Colla, Universita della Tuscia, Viterbo, Italy

Brazilian Seeds *Phaseolus Vulgaris*: Demonstration of Gravitopism and Phototropism Effects on the Germination of Seed in Microgravity (SED)

Expedition: 13
Leading Space Agency: Brazilian Space Agency
Grade Levels: 5–8 (elementary), 9–12 (secondary)

Investigation Description: SED (Brazilian Seeds) was an educational experiment that tested the germination and growth of Brazilian plant seeds under different light conditions. The experiment also featured a control group of similar seeds that was being grown on Earth under similar conditions and for the same time period. The main objective of this experiment was to stimulate and engage students across Brazil in the vast possibilities of space research. Pictures of the growing seeds on ISS were downlinked daily and posted on an Internet site for Brazilian students to see and to compare with their own ground-based seeds.

Commercial Generic Bioprocessing Apparatus Science Insert–01: *C. Elgans* and Seed Germination (CSI-01)

Expeditions: 14, 15

Leading Space Agency: NASA

Grade Levels: 5–8

Impact: This investigation reached 4500 elementary, 5250 secondary, 10 undergraduate, and 5 graduate students as well as 130 schools and 150 teachers.

Investigation Description: Commercial Generic Bioprocessing Apparatus Science Insert–01 (CSI-01) was composed of two educational experiments used by middle school students in the United States and Malaysia. One experiment examined seed germination in microgravity, including gravitropism (plant growth toward gravity) and phototropism (plant growth toward light). The second experiment examined how microgravity affects the model organism, *Caenorhabditis elegans*, a small nematode worm. Thousands of students began participating in the experiments in February 2007.

Student Activity: Each experiment was designed to be easily reproducible in the classroom, providing hands-on experience to the students. The students were able to view the progress of the CSI-01 investigation on ISS via near real-time downlink and the World Wide Web. These experiments have the potential to impact between 7500 and 11,250 students during the first phase of this program, which is designed to be conducted annually.

Principal Investigator: Louis Stodieck, Ph.D., University of Colorado, BioServe Space Technologies, Boulder, Colorado, USA

Education Lead: Stefanie Countryman, BioServe Space Technologies, Boulder, Colorado, USA

Commercial Generic Bioprocessing Apparatus Science Insert–02: Silicate Garden, Seed Germination, Plant Cell Culture and Yeast (CSI-02)

Expeditions: 15, 16, 17, 18

Leading Space Agency: NASA

Grade Levels: 5–8

Impact: This investigation reached 500 elementary, 3500 secondary, 5 undergraduate, and 5 graduate students including 30 schools and 40 teachers.

Investigation Description: CSI-02 was an educational payload designed to interest middle school students in science, technology, engineering and math by participating in near real-time research conducted on board the ISS.

Student Activity: Students observed four experiments through data and imagery downlinked and distributed directly into the classroom via the Internet. The first experiment examined seed germination and plant development in microgravity. The second experiment examined yeast cells' adaptation to the space environment. The third experiment examined plant cell cultures, and the fourth was a silicate garden. The experiments conducted for CSI-02 were designed primarily to meet education objectives; however, to the maximum extent possible, meaningful scientific research was conducted to generate new knowledge of gravity-dependent biological processes and to support future plans for human space exploration. CSI-02 affected more than 15,000 middle school and high school students.

Principal Investigator: Louis Stodieck, Ph.D., University of Colorado, BioServe Space Technologies, Boulder, Colorado, USA

Education Lead: Stefanie Countryman, BioServe Space Technologies, Boulder, Colorado, USA

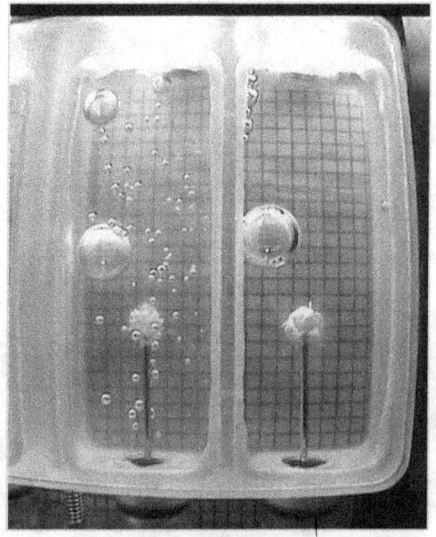

This image shows the nickel sulfate silicate garden grown during Expedition 17. Image courtesy of BioServe Space Technologies, University of Colorado, Boulder, Colorado.

Commercial Generic Bioprocessing Apparatus Science Insert–03: Spiders and Butterflies (CSI-03)

Expeditions: 18, 21/22

Leading Space Agency: NASA

Grade Levels: K–12

Impact: This investigation reached over 95,000 elementary and almost 81,000 secondary students. More than 3000 teachers from 2900 schools were also impacted.

Investigation Description: CSI-03 was one investigation in the CSI program series. The CSI program provided the K–12 community opportunities to use the unique microgravity environment of the International Space Station as part of the regular classroom to encourage learning and interest in science, technology, engineering and math. CSI-03 examined the complete life cycle of the painted lady butterfly, including eating, growing and undergoing metamorphosis in space.

Monarch butterflies in the CSI-03 habitat on board the International Space Station during Expeditions 21/22. Image courtesy of BioServe Space Technologies, Boulder, Colorado.

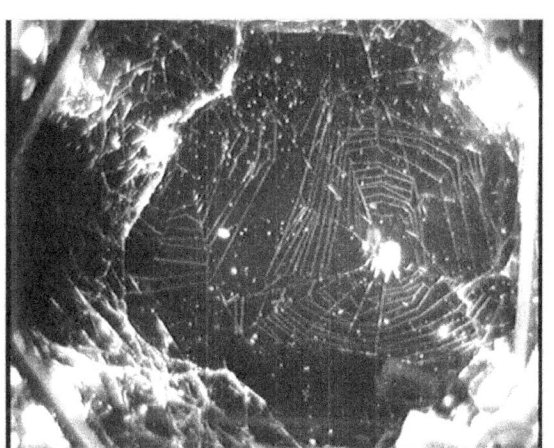

Orb weaving spider with web in CSI-03 habitat onboard the International Space Station during Expedition 18. Image courtesy of BioServe Space Technologies, Boulder, Colorado.

Student Activity: CSI-03 examined the complete life cycle of the *Vanessa cardui*, painted lady butterfly, (egg to butterfly) and the ability of older larvae of a Monarch butterfly species to metamorphosis. Students compared how the complete life cycle of the butterflies in space differs from the life cycle of butterflies on Earth. Students also observed the butterflies on Earth in their classrooms and compared them to imagery from the ISS.

Principal Investigators: Chip Taylor, Monarch Watch, University of Kansas, Lawrence, Kansas; Ken Werner, Gulf Coast Butterflies, Naples, Florida; Louis Stodieck, Ph.D., University of Colorado; and BioServe Space Technologies, Boulder, Colorado

Education Lead: Stefanie Countryman, BioServe Space Technologies, Boulder, Colorado, USA

Table 19 – Number of Teachers by State That Participated in Butterflies in Space

State	Number of Teachers	State	Number of Teachers
Alabama	46	Montana	4
Alaska	17	Nebraska	12
Arizona	41	Nevada	10
Arkansas	34	New Hampshire	14
California	169	New Jersey	93
Colorado	103	New Mexico	18
Connecticut	41	New York	117
Delaware	3	North Carolina	70
Florida	757	North Dakota	2
Georgia	65	Ohio	62
Hawaii	8	Oklahoma	33
Idaho	6	Oregon	11
Illinois	90	Pennsylvania	59
Indiana	36	Rhode Island	11
Iowa	16	South Carolina	16
Kansas	27	South Dakota	17
Kentucky	6	Tennessee	59
Louisiana	19	Texas	360
Maine	24	Utah	17
Maryland	118	Vermont	4
Massachusetts	32	Virginia	78
Michigan	40	Washington	125
Minnesota	22	West Virginia	42
Mississippi	13	Wisconsin	70
Missouri	26	Wyoming	4

Education Payload Operations – Kit C: Plant Growth Chambers (EPO-Kit C)

Expedition: 15
Leading Space Agency: NASA
Grade Levels: K–12

Investigation Description: Education Payload Operations – Kit C Plant Growth Chambers (EPO-Kit C) was an on-orbit plant growth investigation using basil seeds. The still and video imagery acquired was used as part of a national engineering design challenge for students in grades K–12.

Student Activity: Students grew basil seeds (control and flown seeds) to conduct their own science experiments on plant growth using growth chambers created by the students on the ground.

Principal Investigator: Jonathan Neubauer, NASA JSC, Houston, Texas, USA

Education Lead: Matthew Keil, NASA JSC, Houston, Texas, USA

Educational Demonstration of the Effects of Shape and Size on the Recovery of Precompressed Plastic Material (MATI-75)

Expeditions: 17, 18

Leading Space Agency: Roscosmos

Grade Levels: K–8 (elementary), 9–12 (secondary)

Investigation Description: Space experiment MATI-75 was dedicated to the 75th anniversary of MATI — K. E. Tsiolkovsky Russian State Technological University — and the 150th anniversary of the founder of theoretical space science, Konstantin Eduardovich Tsiolkovsky. It was a demonstration of the effect of shape recovery of blanks made from cellular polymeric materials. To demonstrate the effect of shape and size recovery and stabilization when the blanks were exposed to heating and cooling in zero gravity, a plastic piece pre-compressed on the ground.

Student Activity: College, graduate and postgraduate students were given practical skills in staging a space experiment and conducting ground studies of foamed polymer material structure obtained in the ground and space environments.

Principal Investigator: A. P. Petrov, K. E. Tsiolkovsky, Russian State Technological University (MATI), Russia

Website: http://knts.tsniimash.ru/en

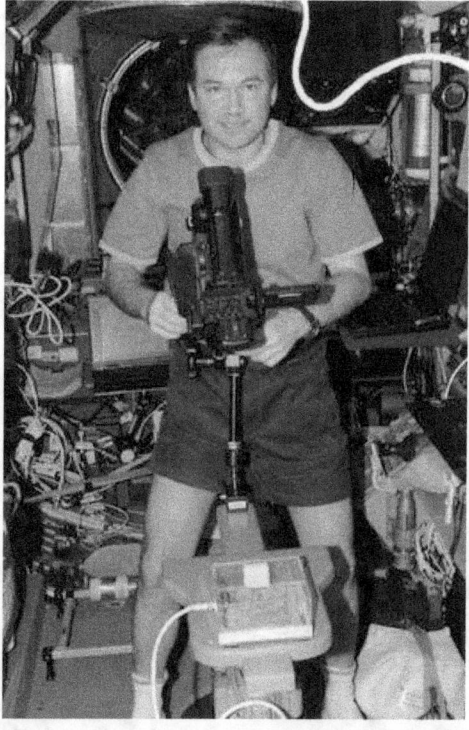

Russian cosmonaut Yu. Lonchakov (onboard engineer on Expedition 18) observes the reconstruction of samples in case No. 1 during the MATI-75 experiment. Image courtesy of Roscosmos.

Case No. 1, which contained flight research samples, before and after shape and dimensions reconstruction onboard the ISS. Image courtesy of Roscosmos.

Farming In Space – Biomass Production System (BPS)

Expedition: 4
Leading Space Agency: NASA
Grade Levels: K–12

Impact: Forty-seven teachers participated in special workshops; four undergraduate pre-service teachers also participated in operating the project. Students and teachers in eight classrooms in Wheeling, West Virginia, participated in Farming in Space.

Investigation Description: The BPS evaluated the performance of a new piece of equipment for growing plants on board ISS. *Brassica rapa* (field mustard) was used as a test species to validate the plant growth system and performance of the equipment. In another experiment, Photosynthesis Experiment and System Testing and Operation (PESTO), wheat (*Triticum aestivum*) was grown in the apparatus as part of a scientific investigation that focused on potential food production in space.

Student Activity: While creating useful technology and science, BPS allowed students in grades K–12 to work as co-investigators on real space research by participating in the Farming in Space program. Farming in Space examined the basic principles and concepts related to plant biology, agricultural production, ecology and the space environment. These activities encouraged curiosity in the sciences while teaching scientific methodology. By performing these activities, the students involved in them learned fundamental aspects of plant biology and gained scientific inquiry skills.

Principal Investigators: Robert Morrow, Orbital Technologies Corporation, Madison, Wisconsin, USA

Education Leads: Laurie Ruberg, Wheeling Jesuit University, Center for Education Technologies, Wheeling, West Virginia, USA.

Oil Emulsion Experiment (OEE)

Expeditions: 13, 14
Leading Space Agency: ESA
Grade Levels: 5–9 (elementary)

Investigation Description: The Oil Emulsion experiment was used to teach students basic principles of fluid physics. Identical experiments were performed on ISS and in the classroom to compare mixing oil and water in microgravity to mixing them on Earth.

Student Activity: A container of oil and colored water was mixed on ISS and another was mixed on Earth so that the reactions of the mixture in gravity (Earth) and microgravity (ISS) could be observed. The fluids' behavior in space was filmed within defined time slots during a two-week period. The data was downlinked, and the results were shown in a special children's program on German public television. The children observed the different kinds of segregation that occurred in space and on Earth during the experiment, and the teacher then explained them. This experiment potentially formed the basis of further physics lessons (concerning weightlessness, density, and other fluid parameters) or even lessons in other scientific areas. The Oil Emulsion experiment was introduced by the German Aerospace Center (DLR) and was a joint effort of the German and the European Space Agencies.

Principal Investigator: Hartmut Ripken, Ph.D., German Aerospace Center, Cologne, Germany

The container with the oil and water mixture that is used in the Oil Emulsion investigation. Image courtesy of ESA.

Seeds in Space (Seeds)

Expeditions: 8, 9
Leading Space Agency: ESA
Grade Levels: 5–9 (elementary)

Investigation Description: The main goal of this student experiment was to involve as many students as possible in an effort to show that science is fun through a plant growth experiment called Seeds in Space. The educational and scientific objective of the experiment was to demonstrate the influence of gravity on the germination and growth of plants to young people (10 to 15 year olds) and others.

Student Activity: By engaging in the comparable on-ground experiment, students learned that science is fun and that the weightless environment of space opens new possibilities. The experiment kit was distributed one month before launch to schools and other entities.

Principal Investigator: Jack van Loon, Free University Amsterdam, Amsterdam, Netherlands

Silkworms in Space – Kinugusa-Kai (Radsilk)

Expeditions: 21/22
Leading Space Agency: JAXA
Grade Levels: K–8 (elementary), 9–12 (secondary)

Impact: There were 20 elementary students, 120 secondary students, 21 schools, and 50 teachers involved in this project.

Investigation Description: Integrated Assessment of Long-term Cosmic Radiation Through Biological Responses of the Silkworm, *Bombyx mori*, in Space (RadSilk) examined the effects of radiation exposure in microgravity on silkworms. The eggs of the silkworm were used as an indicator for monitoring biological responses to long-term cosmic radiation in microgravity.

Student Activity: Students took part in breeding and observing of Cosmos-Silkworms, which had stayed in ISS during their early development.

Principal Investigator: The Kinugasa-Kai Foundation

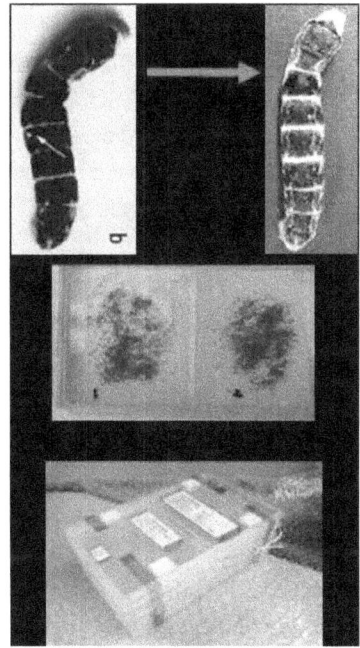

Bombyx mori, silkworms to be used in the RadSilk investigation. Image courtesy of JAXA.

Space-Exposed Experiment Developed for Students (Education-SEEDS)

Expedition: 1

Leading Space Agency: NASA

Grade Levels: K–12

Impact: A total of 750,000 students across the United States participated in the experiments, growing corn and soybean seeds in their classrooms to compare their results with the results from the station and participating in live broadcasts.

Investigation Description: The Investigating Space Exposed Experiment Developed for Students (Education-SEEDS) was a plant seed experiment that tested growth under different light, pressure and microgravity conditions. The experiment was generated from three different types of seeds — a control group, seeds that had been exposed to a Mars-like environment, and seeds that had been exposed to a simulated Mars greenhouse environment. On-orbit videotape and photographic images were taken of plant germination and early growth. Imagery was converted to educational videos to excite and engage students in science and technology and to motivate and provide professional development for educators.

Student Activity: JASON XI — Going to Extremes looked at sea and space through the eyes of modern-day explorers. The *Aquarius* Underwater Laboratory in the Florida Keys was compared to the ISS as a research platform that enables humans to go beyond their physical limitations to explore the unknown. The JASON XI project had two educational components. First, students were selected from across the country and traveled to conduct the experiment with scientists. Second, classroom curricula were created from the experiments that were conducted, and these were used by all participating schools.

Students participating in the "extreme environment" portion of the project assisted scientists during daily one-hour broadcasts from NASA JSC and *Aquarius*. Five live broadcasts occurred each day during the two weeks of the JASON project.

Following completion of JASON XI at the remote locations, packets of seeds that had been exposed to microgravity on the station were provided with the accompanying curriculum to teachers across the country for use in their classrooms. Students grew the seeds the same way the seeds were grown at NASA JSC, on *Aquarius* and on ISS. Using the JASON Web site, the students compared their results to results from NASA JSC, *Aquarius* and the station. The curricula provided by the JASON project met National Science Education Standards, National Geography Standards and National Educational Technology Standards for Students.

Principal Investigator: Howard Levine, Ph.D., NASA KSC, Cape Canaveral, Florida, USA

Students Participating in ISS Investigator Experiments

Many research investigators have enlisted students to help them on their experiments. Some of these experiments were done solely to inspire the next generation of scientists, engineers and explorers. Others involved undergraduate and graduate students as well as postdoctoral fellows, who worked on their research under the guidance of investigators and university professors.

Advanced Diagnostic Ultrasound in Microgravity (ADUM)

Expeditions: 7, 8, 9, 10, 11, 12

Leading Space Agency: NASA

Grade Levels: K–8 (elementary), 9–12 (secondary), College (undergraduate), Graduate (master's, Ph.D., M.D.)

Impact: Ninety K–8 students, 60 9–12 students, 12 undergraduate students and 4 graduate students were involved in this scientific investigation.

Investigation Description: In the ADUM experiment, crewmembers conducted ultrasound exams on one another to determine whether the accuracy of ultrasound is such that it can be used to diagnose certain types of on-orbit injuries and illnesses. Ultrasound may have direct applications related to the evaluation and diagnosis of 250 medical conditions of interest. The experiment was also used to assess the feasibility of using ultrasound to monitor in-flight bone alterations.

The ultrasound device is the only medical imaging device currently available on the International Space Station. This experiment demonstrates the diagnostic accuracy of ultrasound for use in medical contingencies in space and the ability of minimally trained crewmembers to perform ultrasound examinations with guidance from the ground. Crews traveling beyond low Earth orbit will need the telemedicine strategies used in this experiment if a crewmember is injured or becomes ill in space. Earth applications include emergencies and rural care situations.

Student Activity: Ultrasound outreach experiments were conducted at high schools in Michigan and Texas. Students in Michigan developed experimental models that were used on the C9 and used to verify educational methods prior to their deployment on the station.

Description of Teacher Participation and Activities: Ultrasound outreach experiments were conducted at high schools in Michigan and Texas, and the results of the research were published. Student groups in Michigan developed experimental models that were used on the C9 and used to verify educational methods prior to ISS deployment. A high school teacher served as an experimental subject and flew on the C9 to perform microgravity experiments.

Principal Investigator: Scott A. Dulchavsky, M.D., Ph.D., Henry Ford Health System, Detroit, Michigan, USA

Coarsening in Solid Liquid Mixtures-2 (CSLM-2)

Expeditions: 7, 15, 16, 17, 23/24
Leading Space Agency: NASA
Grade Levels: College (undergraduate), Graduate
Impact: One undergraduate, three graduate and one postdoctoral student were involved in this scientific investigation.

Investigation Description: CSLM-2 investigated the coarsening rates of solid particles embedded in a liquid matrix. During this investigation, small particles shrank when they lost atoms to larger particles, which grew (coarsened) within a liquid lead-tin matrix. This study defined the mechanisms and rates of coarsening that govern similar processes that occur in materials, such as turbine blades, dental amalgam fillings, and aluminum alloys.

Student Activity: An undergraduate student analyzed g-jitter data from the ISS to determine its effects on the experiment. She showed that an experiment using a low-volume fraction of a solid is possible. Using her calculations as support, we were able to fly our low-volume fraction solid experiments. The samples were returned in February, and the results of her calculations were published.

Principal Investigator: Peter W. Voorhees, Ph.D., Northwestern University, Evanston, Illinois, USA

Commercial Biomedical Test Module – 2 (CBTM-2)

Expedition: 15
Leading Space Agency: NASA
Grade Levels: College (undergraduate), Graduate (master's, Ph.D., M.D.)
Impact: Eight undergraduate, 21 graduate, and 3 postdoctoral students; 10 schools; and 14 teachers were involved in this investigation.

Investigation Description: This investigation used a validated mouse model to examine the effectiveness of an experimental therapeutic as a possible countermeasure for muscle atrophy.

Student Activity: Undergraduate and graduate students and postdoctoral fellows participated in the scientific process and analysis of space flight samples. In addition, graduate students participated in the launch preparation process of the experiment investigation as well as the sample processing when the space flight samples returned to Earth. Both graduate and undergraduate students participated in the operation activities that occurred while this experiment was on board the ISS.

Principal Investigator: H. Q. Han, Amgen Research, Thousand Oaks, California, USA

Commercial Generic Bioprocessing Apparatus – Antibiotic Production in Space (CGBA-APS)

Expeditions: 2, 4

Leading Space Agency: NASA

Grade Levels: College (undergraduate), Graduate (master's, Ph.D., M.D.)

Impact: 3 undergraduate students, 5 graduate students, and 3 teachers were involved in this investigation.

Investigation Description: Previous studies showed increased antibiotic production during short-duration space flights. The CGBA-APS investigation examined production of actinomycin D, an antibiotic, during long-term exposure to microgravity to determine the mechanism that caused the increased antibiotic production. Once the mechanism was determined, it could be applied to Earth-based pharmaceutical manufacturing techniques.

Student Activity: Students helped design experiment hardware and conduct tests in the investigation.

Principal Investigator: David Klaus, BioServe Space Technologies, University of Colorado, Boulder, Colorado, USA

Crewmember and Crew-Ground Interaction During International Space Station Missions (Interactions)

Expeditions: 2, 3, 4, 5, 7, 8, 9

Leading Space Agency: NASA

Grade Levels: College (undergraduate), Graduate (master's, Ph.D., M.D.)

Impact: One graduate and two postdoctoral students were involved in this scientific investigation.

Investigation Description: Weekly questionnaires were completed to identify and define important interpersonal factors that may affect the performance of the crew and ground-support personnel during International Space Station missions. Results were used to improve the ability of future crewmembers to interact safely and effectively with each other and ground-support personnel. The results may also be used to improve methods of crew selection, training and in-flight support.

Student Activity: Students trained in laboratory activities. One psychiatry resident completed her own project and published her work in two peer-reviewed journals.

Principal Investigator: Nick A. Kanas, M.D., Veterans' Affairs Medical Center and University of California – San Francisco, San Francisco, California, USA

Differentiation of Bone Marrow Macrophages in Space (BONEMAC)

Expedition: 18

Leading Space Agency: NASA

Grade Levels: College (undergraduate), Graduate (master's, Ph.D., M.D.)

Impact: Sixteen undergraduate, 5 graduate, and 3 postdoctoral students were involved in this scientific investigation.

Investigation Description: This experiment investigates how long-term exposure to microgravity, such as that experienced on missions to the moon and Mars, affects the production of cells critical to the human immune system. Exposure to microgravity interferes with several important immune system and blood cell production processes. Disruption of the development of white blood cells (cells which are the first line of defense against new pathogens) is of particular concern for future space flights to the moon and Mars. The BONEMAC experiment investigates the effects of microgravity on macrophage production and development using rodent bone marrow. The study also examines how bone loss during space flight may adversely impact blood cell production.

Student Activity: Students assisted with sample preparation, observation, laboratory assays and preparation.

Principal Investigator: Stephen K. Chapes, Kansas State University, Manhattan, Kansas, USA

Fluid Merging Viscosity Measurement (FMVM)

Expeditions: 9, 11

Leading Space Agency: NASA

Grade Levels: College (undergraduate); Graduate (master's, Ph.D., M.D.)

Impact: One graduate student was involved in this scientific investigation.

Investigation Description: This experiment was designed to test a new method for measuring the viscosity of high-viscosity materials by measuring the time it takes two nearly free-floating drops of a liquid to merge. The materials used were of known viscosities (corn syrup, glycerin and silicone oil) so that the accuracy of the fluid merging test could be compared to the methods used on Earth. The FMVM experiment led investigators to a greater understanding of how glass is formed from melted lunar soil. It also has led to a better understanding of liquid phase sintering processes for in-space fabrication methods that can be used to construct surface habitats.

Student Activity: Students assisted with sample preparation, observation, laboratory assays and preparation.

Principal Investigator: Edwin C. Ethridge, NASA Marshall Space Flight Center, Huntsville, Alabama, USA

International Space Station Zero-Propellant Maneuver (ZPM) Demonstration

Expeditions: 14, 15, 16, 17, 18, 19/20, 21/22, 23/24, 25/26
Leading Space Agency: NASA
Grade Levels: College (undergraduate), Graduate (master's, Ph.D., M.D.)
Impact: Three graduate students were involved in this scientific investigation.

Investigation Description: The International Space Station ZPM demonstration showed for the first time new technology that rotated the station by not expending on-orbit propellant. The capability to rotate the station was necessary for assembly and operation; e.g., reorientation of the ISS to allow docking of the space shuttle and Russian resupply vehicles. Rotational maneuvers were performed using the station's thrusters, which consumed precious propellant. The goal of the ZPM demonstration was to reduce the cost of operating the ISS and to provide a backup to the thrusters.

Student Activity: The flight demonstration was performed using the ZPM trajectory, which developed as a master's thesis by Sagar Bhatt in the Department of Computational and Applied Math at Rice University; the investigation was supervised jointly by Professor Yin Zhang of Rice and Dr. Nazareth Bedrossian of the Charles Stark Draper Laboratory. The thesis was performed under a Draper Laboratory Fellow (DLF) appointment for Sagar. The DLF program develops low-cost high-impact innovation for NASA by having the Fellows perform a thesis on a NASA related problem. Under this program, Sagar was able to have extensive interactions with JSC staff working on the International Space Station where he participated in meetings, presentations, and problem solving.

Description of Teacher Participation and Activities: Dr. Nazareth Bedrossian (Charles Stark Draper Laboratory) was the principal investigator on the ZPM project, while Professor Yin Zhang (Rice University) was the academic advisor. Dr. Bedrossian, who originated the ZPM concept, provided direction in formulating the problem for the International Space Station and interaction with NASA personnel. Professor Zhang provided direction in solving the optimization problem.

Principal Investigator: Nazareth Bedrossian, Ph.D., Charles Stark Draper Laboratory Inc., Houston, Texas, USA

Naval Postgraduate School (NPS) Professor of Mechanical and Astronautical Engineering I. Michael Ross briefs NPS Space Systems Academic Group students on NASA's use of his optimal control software to maneuver the International Space Station cost-free, without the need to use thrusters and expend valuable fuel. U.S. Navy photo by Javier Chagoya.

Intravenous Fluid Generation Experiment (IVGEN)

Expeditions: 23/24

Leading Space Agency: NASA

Grade Levels: College (undergraduate); Graduate (master's, Ph.D., M.D.)

Impact: Four students in grades 9–12 and 3 undergraduate students were involved in this investigation.

Investigation Description: IVGEN demonstrates the ability to purify water to the standards required for intravenous administration, and then mix the water with salt crystals to produce normal saline. The hardware is a prototype that will allow flight surgeons more options to treat ill or injured crewmembers during future long-duration exploration missions.

Student Activity: Students performed a variety of tests to measure both the efficiency of purification and the hydrodynamic characteristics of a deionizing (DI) resin-packed bed filter cartridge used to purify water. The project was performed in support of the IV-Gen Project, whereby tap water will be purified to sterile water and mixed with salt to produce a normal saline intravenous solution. Students prepared resin cartridges and integrated them into the flow loop, conducted tests over a wide range of system temperatures, and analyzed and plotted data. A student also conducted literature surveys on purification standards for the United States Pharmacopeia.

Principal Investigator: John McQuillen, NASA Glenn Research Center, Cleveland, Ohio, USA

Italian Foam (I-Foam)

Expeditions: 27/28

Leading Space Agency: Italian Space Agency (ASI)

Grade Levels: 9–12 (secondary); College (undergraduate); Graduate (master's, Ph.D., M.D.)

Impact: One undergraduate student and 2 graduate students were involved in this scientific investigation.

Investigation Description: The I-Foam experiment evaluated the recovery of shape memory epoxy foam in microgravity. The foam was obtained by solid-state foam on the ground, which consisted of various geometric complexities that were shaped on the ground. This investigation studied the shape memory properties required to manufacture a new concept actuator (a device that transforms energy to other forms of energy).

Student Activity: Undergraduate and graduate students were involved in the investigation. They worked mainly in the laboratory, performing the ground experiments and collecting results for their degree thesis. Loredana Santo and Fabrizio Quadrini are associate professors at the University of Rome Tor Vergata in Italy. The students' fields of study are technology and manufacturing systems. They were involved in each activity of the investigation, planning ground and flight experiments, performing ground experiments, and analyzing the results analysis.

Principal Investigator: Professor Loredana Santo, University of Rome Tor Vergata, Rome, Italy

Mice Drawer System (MDS)

Expeditions: 19/20, 21/22
Leading Space Agency: ASI
Grade Levels: 9–12 (secondary)
Impact: Three undergraduate, 10 graduate, and 9 postdoctoral students were involved in this scientific investigation.

Investigation Description: MDS is an ASI investigation that uses a validated mouse model to investigate the genetic mechanisms underlying bone mass loss in microgravity. Research conducted with the MDS is an analog to the Human Research Program, the objective of which is to extend the human presence safely beyond low Earth orbit. The Mice Drawer System (MDS) is able to support mice onboard the International Space Station during long-duration exploration missions (from 100 to 150 days) by providing living space, food, water, ventilation and lighting. Mice can be accommodated either individually (to a maximum of six) or in groups (four pairs). Osteoporosis is a debilitating disease that afflicts millions of people worldwide. One of the physiological changes experienced by crewmembers during space flight is the accelerated loss of bone mass due to the lack of gravitational loading on the skeleton. This bone loss experienced by crewmembers is similar to osteoporosis in the elderly population. MDS investigated the effects of unloading on transgenic (foreign gene that has been inserted into its genome to exhibit a particular trait) mice with the Osteoblast Stimulating Factor-1, OSF-1, a growth and differentiation factor, and studied the genetic mechanisms underlying the bone mass pathophysiology. MDS tested the hypothesis that mice with an increased bone density are likely to be more protected from osteoporosis, when the increased bone mass is a direct effect of a gene involved in skeletogenesis (skeleton formation).

Student Activity: Many Ph.D. and postdoctoral students were involved in the planning, preparation, execution and data analysis of the MDS experiment. The purpose of the experiment was to investigate changes that occurred in different mouse tissues in microgravity conditions; the experiment was launched to the ISS and spent 91 days in space, making it the first long-duration (three months) animal experiment on the International Space Station. To obtain as much information as possible on microgravity-induced tissue modifications in animals, a Tissue Sharing Program was created among 19 groups from six countries.

Principal Investigator: Ranieri Cancedda, M.D., University of Genoa (Unige) and National Cancer Research Institute, Genoa, Italy

Mouse Antigen-Specific CD4+ T Cell Priming and Memory Response During Spaceflight (Mouse Immunology)

Expeditions: 23/24

Leading Space Agency: NASA

Grade Levels: 9–12 (secondary), College (undergraduate), Graduate (master's, Ph.D., M.D.)

Impact: Three secondary, one undergraduate and one graduate student were involved in this scientific investigation.

Investigation Description: The Mouse Immunology investigation studies specific mechanisms of immune system activation and whether immune system cells exposed to challenges before flight retain the "memory" to fight challenges during space flight. Explorers on future long-duration space missions may require pre-flight vaccinations or other precautions to prevent infection during space travel if immune memory is not retained.

Student Activity: Students participated in laboratory procedures and basic laboratory duties.

Principal Investigator: Millie Hughes-Fulford, Ph.D., University of California – San Francisco, San Francisco, California, USA

National Laboratory Pathfinder – Cells – *Jatropha* Biofuels (NLP-Cells-3, 4, 6, 7)

Expeditions: 21/22, 23/24, 25/26, 27/28

Leading Space Agency: NASA

Grade Levels: College (undergraduate); Graduate (master's, Ph.D., M.D.)

Impact: Five undergraduate students, seven graduate students and one postdoctoral student were involved in this scientific investigation.

Investigation Description: National Lab Pathfinder-Cells-*Jatropha* Biofuels (NLP-Cells-*Jatropha* Biofuels) assesses the effects of microgravity on the formation, establishment and multiplication of undifferentiated cells of the *Jatropha (Jatropha curcas)*, a biofuel plant, using different tissues as explant sources from different genotypes of *Jatropha*. Specific goals include the evaluation of changes in cell structure, growth and development, genetic changes, and differential gene expression. Postflight analysis identifies significant changes that occur in microgravity that could contribute to acceleration of the breeding and genetic improvement processes for the development of new cultivars of this biofuel plant.

Student Activity: Support in preparing materials for experiment, such as preparing the medium; organizing data; labeling; photographing samples; collecting pre- and postflight data; and helping with integration, analyzes, and related activities.

Principal Investigator: Wagner Vendrame, Ph.D., University of Florida, Homestead, Florida, USA

Passive Observatories for Experimental Microbial Systems (POEMS)

Expeditions: 13, 14

Leading Space Agency: NASA

Grade Levels: College (undergraduate), Graduate (master's, Ph.D., M.D.)

Impact: Two 9–12 students and 5 undergraduate students were involved in this scientific investigation.

Investigation Description: This investigation involved demonstrating a passive system for growing microbial cultures in space and observing genetic changes that occur as a result of living and growing in space. POEMS helped scientists understand the growth, ecology, and performance of diverse assemblages of microorganisms in space required to maintain human health and bioregenerative function in support of advanced life support systems for future exploration vehicles.

Student Activity: The design, development, testing, and engineering of the Biological Research in Canisters POEMS (BRIC-POEMS) flight hardware as well as payload verification testing and science verification testing activities were supported by undergraduate interns at KSC. Student interns assisted the payload engineering team in developing computer-aided design drawings of the BRIC-POEMS hardware and completing materials compatibility and biological safety testing for the hardware. The interns assisted the science investigation team in ground experiments to verify the performance of the BRIC-POEMS hardware for support of science objectives and to optimize test procedures.

Principal Investigator: Michael Roberts, Ph.D., Dynamac Corporation, Cape Canaveral, Florida, USA

Pore Formation and Mobility During Controlled Directional Solidification in a Microgravity Environment (PFMI)

Expeditions: 5, 7, 8, 13

Leading Space Agency: NASA

Grade Levels: College (undergraduate), Graduate (master's, Ph.D., M.D.)

Impact: Two graduate students were involved in this scientific investigation.

Investigation Description: Using a transparent model material, the PFMI experiments studied the fundamental phenomena responsible for the formation of certain classes of defects in metal castings. Investigators examined the physical principles that control the occurrence of defects in manufacturing on Earth to develop methods to reduce flaws, defects or wasted material.

Student Activity: The results of PFMI experiments contributed to a doctoral degree and a master's degree. Plane Front Dynamics and Pattern Formation in Diffusion Controlled Directional Solidification of Alloys, by Louise Strutzenberg, is a dissertation that partially fulfills the requirements for a Doctor of Philosophy degree in materials science and engineering. Experimental Study of Wormhole Growth and Evolution in Small Cylindrical Channels during Directional Solidification, by Matthew C. Cox, is a dissertation that partially fulfills the requirements for a Master of Science degree in mechanical engineering.

Principal Investigator: Richard Grugel, Ph.D., NASA Marshall Space Flight Center, Huntsville, Alabama, USA

Sleep-Wake Actigraphy and Light Exposure During Spaceflight-Long (Sleep-Long) and Sleep-Wake Actigraphy and Light Exposure During Spaceflight-Short (Sleep-Short)

Expeditions: 14, 15, 16, 17, 18, 19/20, 21/22, 23/24, 25/26

Leading Space Agency: NASA

Grade Levels: College (undergraduate), Graduate (master's, Ph.D., M.D.)

Impact: Three graduate students and two postdoctoral students were involved in this scientific investigation.

Investigation Description: Sleep-Long examines the effects of space flight and ambient light exposure on the sleep-wake cycles of crewmembers during long-duration stays on board the International Space Station. Sleep-Short examines how space flight affects crewmembers' sleep patterns during Space Shuttle missions.

Principal Investigators: Laura K. Barger, Ph.D., Brigham and Women's Hospital, Harvard Medical School, Boston, Massachusetts; Charles A. Czeisler, M.D., Ph.D., Brigham and Women's Hospital, Harvard Medical School, Boston, Massachusetts, USA

Space Flight Induced Reactivation of Latent Epstein-Barr Virus (Epstein-Barr)

Expeditions: 5, 6, 11, 12, 13, 14, 15, 16, 17
Leading Space Agency: NASA
Grade Levels: 9–12 (secondary)

Impact: Two 9-12 students were involved in this scientific investigation.

Investigation Description: The Space Flight Induced Reactivation of Latent Epstein-Barr Virus (Epstein-Barr) experiment investigated changes in the human immune function using blood and urine samples collected before and after space flight. The study provided insight into possible countermeasures to prevent the potential development of infectious illness in crewmembers during flight. As space-flight mission durations increase, the potential for development of infectious illnesses in crewmembers during flight also increases. This is especially true with latent viruses (viruses such as herpes viruses that lie dormant in cells and cause cold sores) and infections caused by these viruses that are not mitigated by a quarantine period. An example of a latent infection is the Epstein-Barr virus, of which approximately 90 percent of the adult population is infected. Stress and other acute or chronic events reactivate this virus after latency, which results in increased virus replication. Epstein-Barr assessed the immune system response to this virus using blood and urine samples collected before and after space flight.

Student Activity: Students participated in space studies at the University of Texas Medical Branch through the National Space Biomedical Research Institute Summer Student Program. They spent a summer in various laboratories and presented their research in poster forums.

Principal Investigator: Raymond Stowe, Ph.D., Microgen Laboratories, La Marque, Texas, USA

Student Access to Space (Part of Protein Crystal Growth-Enhanced Gaseous Nitrogen Dewar) (PCG-EGN)

Expeditions: 0–2, 4

Leading Space Agency: NASA

Grade Levels: College (undergraduate), Graduate (master's, Ph.D., M.D.)

Impact: Approximately 58,000 students and almost 1200 teachers have been involved in this program. Of those, more than 420 students and 260 teachers have been involved in the Student Flight Sample.

Investigation Description: The primary objective of the Protein Crystal Growth – Enhanced Gaseous Nitrogen (PCG-EGN) experiment was to provide a low-cost, simple platform for the production of a large number of protein crystals in space. Prior to launch, scientists placed the samples in the EGN dewar (a thermos-like container that had an absorbent inner liner saturated with liquid nitrogen). Launched aboard the shuttle, the sample dewars were transferred to ISS for processing. After about 10 days, when the nitrogen had evaporated and the samples had thawed, the biological solutions began to form crystals. When returned to Earth, the crystals were analyzed for internal quality, three-dimensional structure and factors that influenced crystal growth.

Student Activity: Classroom protein crystal growth kits were distributed to the classrooms, with each kit supporting three classroom experiments that took seven to 10 days to complete. The experiments compared eight different conditions for crystal growth, and students counted the number of crystals grown and measured their size over time for each growing condition. By doing this, students were taught the process of making observations, recording data and preparing experiment reports. Students also competed to participate in the Protein Crystals in Space Program, where they attended a Student Flight Sample Workshop. In the workshops, which were held before each experiment's flight to the International Space Station, students and teachers worked beside scientists from the University of California, Irvine, and NASA. They prepared, froze and sealed the protein solutions in small tubes. This activity taught students how to perform crystallization experiments in the classroom and on the ISS. The students prepared and loaded actual flight samples into the flight EGN-dewar facility, were present at launch, and received their samples after flight. On Earth, students viewed photos of some crystals grown during the NASA workshops.

Principal Investigator: Alexander McPherson, Ph.D., University of California, Irvine; Irvine; California, USA

Students Participating in SS Engineering Education: Hardware Development

These activities typically involved high school and college students who developed hardware to support the International Space Station Program.

Filming of Space Robot "Jitter" Assembled Out of LEGOS (Konstructor)

Expedition: 4

Leading Space Agency: Roscosmos

Grade Levels: K–8 (elementary), 9–12 (secondary)

Investigation Description: This student activity involved the filming of space robot Jitter, which had been assembled out of LEGO components.

Student Activity: Scripted activities were filmed in a habitable module using a DVCAM video camera, and a videocassette of the film was returned to Earth. The video was used as a teaching aid to inspire students studying science, engineering, technology and mathematics.

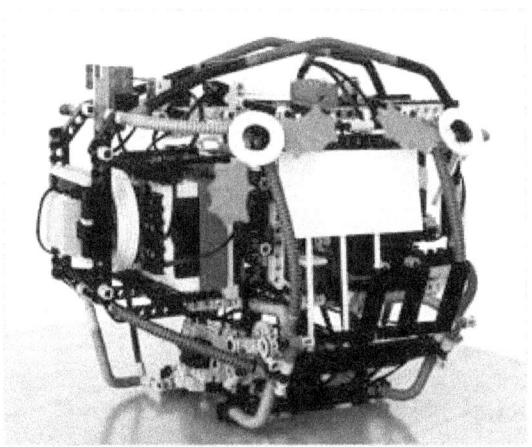

Jitter Robot. Image courtesy of Roscosmos.

Kolibry

Expedition: 4

Leading Space Agency: Roscosmos

Grade Levels: College (undergraduate), Graduate (master's, Ph.D., M.D.)

Investigation Description: The goal of the Kolibry microsatellite launch was to allow pupils to operate microsatellites under the direct guidance of scientists. Students were responsible for implementing projects dealing with ballistics, radio engineering, computer science and thermal physics. They analyzed and observed the satellite manufacturing process and analyzed the scientific information transmitted from the space satellite. The operation contributed to the further improvement of the secondary and high education system, and it will continue to make it possible to find the best solutions to the problems of training future specialists for the space industry.

Educational Demonstrations and Activities

In addition to the numerous International Space Station experiments that students supported, several other educational demonstrations and activities were performed to be used as teaching aids, to supply educational resource materials, or simply to provide additional mechanisms to inspire students.

Demonstration of Mass and Weight of Objects and Action of Reactive Forces in Microgravity (Education-SA)

Expedition: 4

Leading Space Agency: South African Space Agency

Grade Levels: K–8 (elementary), 9–12 (secondary)

Investigation Description: Education-SA provided schoolchildren a visual demonstration of the difference between an object's mass and weight, the action of reactive force, the inertia of resting and moving bodies in zero gravity, and the behavior of an object in zero gravity and in terrestrial gravity. Video footage taken during the experiment will be used to improve visual aids to make school teaching in South Africa more efficient and encourage schoolchildren to take an interest in space exploration.

Electronic-Learning (E-Learning)

Expedition: 14

Leading Space Agency: ESA

Grade Levels: K–8 (elementary), 9–12 (secondary)

Investigation Description: E-Learning involved interactive, real-time lectures given by an ESA astronaut on the International Space Station. These lectures to universities across Europe were a unique opportunity for university students and helped to inspire the next generation of space explorers.

Student Activity: The E-learning session was scheduled as an "Earth-based" lecture during which there was a live audio-video link with ESA astronaut Thomas Reiter, who was on board the ISS. The lecture was presented to European university students following the EuMAS Master's Program in Aeronautics and Space by Dr. Hubertus Thomas of the Max-Planck Institute in Garching, Germany. The topic of the lecture was plasma crystals and complex plasmas, with reference to the Plasma Crystal Research (PK-3+ experiment) that was to be performed on the station during the Astrolab mission.

Education — How Solar Cells Work (Education-Solar Cells)

Expedition: 13
Leading Space Agency: NASA
Grade Levels: K–8 (elementary), 9–12 (secondary)

Investigation Description: The astronaut discussed in detail how solar cells work and how they provided energy; the activity was videotaped and used in classroom lectures. Students developed a basic understanding of electricity and power and the variables that affect the operation of solar panels. They also learned about the power requirements of the International Space Station and how solar arrays supply the necessary power as well as the power requirements of their own homes and how solar power could supply that power.

Principal Investigator: Astronaut Christopher J. Ferguson, NASA JSC, Houston, Texas, USA

Greenhouse in Space (ESA-EPO-Greenhouse)

Expeditions: 25/26
Leading Space Agency: ESA
Grade Levels: K–8 (elementary), 9–12 (secondary)

Impact: Sixteen thousand students, 800 schools, and 800 teachers were involved in this investigation.

Investigation Description: Students set up miniature greenhouses (supplied by ESA) and planted their seeds (Arabidopsis thaliana) at the same time that crewmember Paolo Nespoli planted seeds on the International Space Station. They studied the development of the seeds and investigated how variables could be changed to achieve successful germination. The 10-week project covered the full life cycle of a flowering plant.

Student Activity: The space experiment developed mold after three weeks and had to be trashed for safety reasons. Despite this, many schools continued the project and shared their results and observations on a dedicated Facebook page. This was linked to an online lesson on a website that was developed to guide teachers and students through the project. The project made students aware of the work cosmonauts do on the space station and the unique conditions of microgravity in space. The crew of Mars500 started their space experiment greenhouse at the same time, and this became the focus of the continuing "space" experiment.

The live event linked four countries and approximately 1000 students with the space station on the same day. Recordings of the space experiment are available online for all other schools to use at any time.

Description of Teacher Participation and Activities: Teachers guided students on the project. Some teachers used the project to help students develop their own miniature projects and introduce aspects of the curriculum biology. In some countries, the project was used as an interdisciplinary means of helping students improve their written English. Schools were invited to send their project results to the education team — results will be published online shortly.

Principal Investigator: Ms. Shamim Hartevelt, ESTEC, Netherlands

Students involved in constructing greenhouses and planting seeds. Image courtesy of ESA.

Kibo Kids Tour

Expeditions: 19/20
Leading Space Agency: JAXA
Grade Levels: 5–8 (elementary)

Investigation Description: The Kibo Kids Tour introduced on-orbit Japanese utilization activities in the Kibo module of the International Space Station to the children of Japan. A detailed description of the Kibo module was given along with descriptions of its facilities and experiments. Recorded video was downlinked to Earth after the activity; the crew also videotaped the operations.

The video was then used in developing education curriculum support materials for distribution to educators.

Science of Opportunity (Saturday Morning Science)

Expedition: 6
Leading Space Agency: NASA
Grade Levels: K–8 (elementary), 9–12 (secondary), College (undergraduate), Graduate (master's, Ph.D., M.D.)

Investigation Description: Donald (Don) Pettit, Expedition 6 NASA International Space Station Science Officer, used his free time, usually Saturday mornings, while living aboard the station to shed the light of science on a variety of subjects for students of all ages. These demonstrations were chronicled and dubbed "Saturday Morning Science." The experiments used simple materials that would not affect station operations. During Expedition 6, several scientific principles were demonstrated through Saturday Morning Science. The value of this science is the ability to provide observation-based insights for the reduced gravity environment.

Close-up view of earplugs strung together by NASA International Space Station science officer, Donald (Don) Pettit, to create a sprouter for basil and tomato seeds used in the Growing Plants in Zero-G investigation of Saturday Morning Science during Expedition 6. NASA image ISS006E20853.

Water bubble injected with many air bubbles on board the station during Expedition 6. Image courtesy of NASA.

Some of the many experiments that were performed are described below:

- **Noctilucent Cloud Observations in the Southern Hemisphere** – Noctilucent clouds (clouds that occur in the polar regions in the upper atmosphere, about 80 kilometers up) appear as a thin but distinct cloud layer well above the visible part of the atmosphere. ISS crewmembers have an excellent vantage point for observing these phenomena. Pettit's observations included spacecraft position, date and time, and approximate viewing direction. These observations were compiled into a data set that laid the groundwork for the station's participation in International Polar Year in 2007.

- **Water Observations** – An analog to planetary accretion processes used a plastic bag, sugar, tea grains and water. The particles (e.g., tea leaves) were suspended in water, then manipulated and photographed. The "planetary" accretion (increase in the mass of an object by the collection of surrounding "interstellar gases and objects" through gravity) of the particles in microgravity was observed.

- **Studying Water Films** – In microgravity, thin films were surprisingly robust and could withstand numerous mechanical durability tests without breaking. Blowing on the film created ripples that quickly dampened when the perturbations ceased. Oscillating the loop through tens of centimeters with a period of about 2 seconds distorted the film with patterns like those seen in a soft rubber membrane when it is driven by a sound oscillator.

Principal Investigator: Astronaut Donald (Don) Pettit, Ph.D., NASA JSC, Houston, Texas, USA

Taste in Space

Expeditions: 23/24

Leading Space Agency: ESA

Curriculum Grade Levels: K–8 (elementary), 9–12 (secondary)

Investigation Description: Taste in Space demonstrated the way the sense of taste is affected in microgravity.

Description of Student Participation and Activities: Students were able to follow the blind tasting of crewmember foods by crew on board the ISS and on the ground, performed similar experiments in the classroom and compared the results. They formed conclusions based on the physiological effects of microgravity on the perception of taste.

Principal Investigator: Shamim Hartevelt, ESTEC, Netherlands

Cultural Activities

These educational activities provided students a cultural experience and linked science with the humanities.

DREAMTiME (DREAMTiME)

Expedition: 3

Leading Space Agency: NASA

Grade Levels: K–8 (elementary), 9–12 (secondary), College (undergraduate), Graduate (master's, Ph.D., M.D.)

Investigation Description: DREAMTiME supplied HDTV video cameras and obtained high-quality video footage of activities on the station for commercial, historical, educational, training and public-interest use. The core of DREAMTiME is an HD camera and recorder. DREAMTiME was used on the International Space Station to provide enhanced images and audio to ground-based observers. DREAMTiME helped move NASA into the new millennium through a multimedia revolution, digitizing NASA's archives to share globally via the Web for generations and creating world-class film, documentary and television programming.

Principal Investigator: Ben Mason, Dreamtime Holdings Inc., Moffett Field, California, USA

Get Fit for Space Challenge with Bob Thirsk (Get Fit for Space)

Expeditions: 19/20, 21/22

Leading Space Agency: CSA

Grade Levels: K–8 (elementary), 9–12 (secondary)

Investigation Description: The Get Fit for Space Challenge with Bob Thirsk (Get Fit for Space) was a CSA-sponsored activity that invited Canadian citizens to celebrate the historic mission of Canada's first Expedition along with crewmember Dr. Bob Thirsk and to promote healthy living among Canadian citizens by tracking fitness data using a pedometer, allowing the citizens to interact with Canada's space program via innovative multimedia.

Principal Investigator: Astronaut Robert Thirsk, Ph.D., CSA, St. Hubert, Quebec, Canada

Suit Satellite (RadioSkaf)

Expeditions: 12, 27/28, Ongoing
Leading Space Agency: Roscosmos
Grade Levels: K–8 (elementary), 9–12 (secondary)

Investigation Description: SuitSat-1 uses a decommissioned Orlan spacesuit equipped with a ham radio transmitter and a compact disk containing messages and images from students around the world. After being released during an EVA, the suit transmitted a ham radio signal for approximately six days, and a month later reentered the atmosphere and burned up. Students and hobbyists around the world tuned in to the signal to identify the transmitted words and image. This investigation was conducted to inspire the next generation of explorers, but also assisted in bridging the cultural gap between people around the globe.

Student Activity: Students, teachers, scout troops, ham radio operators and the general public were encouraged to track the signals from SuitSat-1 and listen to the messages. The voices and images coming from SuitSat-1 were collected from students around the world. As SuitSat-1 floated in space, it transmitted voice messages from students in Russian, Japanese, Spanish, German, French, and English. These messages contained a special word, which was copied in different languages from SuitSat-1 and submitted to the ARISS team. SuitSat-1 downlinked images using a series of audio tones, similar to those of a computer modem, using the ham radio picture standard of Slow Scan Television and of similar quality as received on cell phones.

SuitSat-1 created using a decommissioned Russian Orlan spacesuit that is outfitted with transmission hardware will be released by hand during an EVA.
NASA image ISS012E15666

Principal Investigator: Alexander P. Alexandrov, S.P. Korolev RSC "Energia," Russia

Education Website: http://radioskaf.ru/en/

SuitSat-1 in orbit after its release from the station during Expedition 12.
NASA image ISS012E16899

Russian cosmonaut Dmitri Kondrat'ev (onboard engineer on Expedition 26/27) performs a test checkout of the satellite ARISSat-1/Kedr.

Expedition 26 crewmembers with the satellite ARISSat-1/Kedr.

Launch of the second satellite SuitSat-2 (ARISSat-1/Kedr) by the Russian cosmonauts Sergej Volkov and Alexander Samokutyaev (Expeditions 27/28).

Summary of International Space Station Education Accomplishments

Table 20 – Student, Teacher and School Summary of ISS Education Accomplishments

Investigations	Number of Students				Schools	Teachers
	K-12	Undergraduate	Graduate	Postdoctoral		
Student-Developed Investigations	3300	21			20	
Education Competitions						
Students Performing Classroom Versions of ISS Investigations	940,513	29	20		3089	3287
Students Participating in ISS Investigators Experiments	58,161	58	68	21	31	1238
Students Participating in ISS Engineering Education: Hardware Development						
Educational Demonstrations and Activities	16,000				800	800
Cultural Activities						
Totals	~1,020,000	~100	~90	~20	~3900	~5300

Appendix
International Space Station Missions

Table 21 – International Space Station Expeditions and Crewmembers

Expedition		Dates	ISS Crewmembers			
0		11/98 – 11/00				
1		11/00 – 03/01		**William M. Shepherd** Yuri P. Gidzenko Sergei K. Krikalev		
2		03/01 – 08/01		**Yury V. Usachev** James S. Voss Susan J. Helms		
3		08/01 – 12/01		**Frank L. Culbertson** Vladimir N. Dezhurov Mikhail Tyurin		
4		12/01 – 06/02		**Yury I. Onufrienko** Daniel W. Bursch Carl E. Walz		
5		06/02 – 12/02		**Valery G. Korzun** Peggy A. Whitson Sergei Y. Treschev		
6		11/02 – 05/03		**Kenneth D. Bowersox** Donald R. Pettit Nikolai M. Budarin		
7		04/03 – 10/03		**Yuri I. Malenchenko** Edward T. Lu		
8		10/03 – 04/04		**C. Michael Foale** Alexander Y. Kaleri		
9		04/04 – 10/04		**Gennady I. Padalka** E. Michael Fincke		
10		10/04 – 4/05		**Leroy Chiao** Salizhan S. Sharipov		
11		04/05 – 10/05		**Sergei K. Krikalev** John L. Phillips		
12		10/05 – 04/06		**William S. McArthur, Jr.** Valery I. Tokarev		
13		03/06 – 09/06		**Pavel Vinogradov** Jeffrey N. Williams Thomas A. Reiter		
14		09/06 – 04/07		**Michael E. Lopez-Alegria** Mikhail V. Tyurin Thomas A. Reiter		Sunita L. Williams
15		04/07 – 10/07		**Fyodor N. Yurchikhin** Oleg V. Kotov Sunita L. Williams		Clayton C. Anderson
16		10/07 – 04/08		**Peggy A. Whitson** Yuri I. Malenchenko Clayton C. Anderson		Daniel M. Tani Léopold Eyharts Garrett E. Reisman

Expedition	Dates	ISS Crewmembers			
17	04/08 – 10/08		**Sergei A. Volkov** Oleg D. Kononenko Garrett E. Reisman		Gregory E. Chamitoff
18	10/08 – 03/09		**E. Michael Fincke** Yury V. Lonchakov Gregory E. Chamitoff		Sandra H. Magnus Koichi Wakata
19	03/09 – 05/09		**Gennady I. Padalka** Michael R. Barratt Koichi Wakata		
20	05/09 – 10/09		**Gennady I. Padalka** Michael R. Barratt Koichi Wakata Timothy L. Kopra		Roman Y. Romanenko Robert B. Thirsk Frank De Winne Nicole P. Stott
21	10/09 – 11/09		**Frank De Winne** Roman Y. Romanenko Robert B. Thirsk		Nicole P. Stott Jeffrey N. Williams Maxim Suraev
22	11/09 – 03/10		**Jeffrey N. Williams** Maxim Suraev Oleg V. Kotov		Soichi Noguchi Timothy J. Creamer
23	03/10 – 05/10		**Oleg V. Kotov** Soichi Noguchi Timothy J. Creamer		Alexander Skvortsov Tracy Caldwell Dyson Mikhail B. Kornienko
24	06/10 – 09/10		**Alexander Skvortsov** Tracy Caldwell Dyson Mikhail B. Kornienko		Shannon Walker Douglas H. Wheelock Fyodor N. Yurchikhin
25	09/10 – 11/10		**Douglas H. Wheelock** Fyodor N. Yurchikhin Shannon Walker		Alexander Y. Kaleri Scott J. Kelly Oleg I. Skripochka
26	11/10 – 03/11		**Scott J. Kelly** Alexander Y. Kaleri Oleg I. Skripochka		Catherine G. Coleman Dmitry Kondratyev Paolo A. Nespoli
27	03/11 – 05/11		**Dmitry Kondratyev** Catherine G. Coleman Paolo A. Nespoli		Andrey I. Borisenko Ronald J. Garan, Jr. Alexander M. Samokutyaev
28	05/11 – 09/11		**Andrey I. Borisenko** Ronald J. Garan, Jr. Alexander M. Samokutyaev		Sergei A. Volkov Michael E. Fossum Satoshi Furukawa
29	09/11 – 11/11		**Michael E. Fossum** Satoshi Furukawa Sergei A. Volkov		Anton N. Shkaplerov Anatoly A. Ivanishin Daniel C. Burbank
30	11/11 – 04/12		**Daniel C. Burbank** Anton N. Shkaplerov Anatoly A. Ivanishin		Oleg D. Kononenko André Kuipers Donald R. Pettit
31	04/12 – 06/12		**Oleg D. Kononenko** André Kuipers Donald R. Pettit		Joseph M. Acaba Gennady I. Padalka Sergei N. Revin
32	07/12 – 09/12		**Gennady I. Padalka** Joseph M. Acaba Sergei N. Revin		Sunita L. Williams Yuri I. Malenchenko Akihiko Hoshide

Expedition		Dates	ISS Crewmembers			
33		09/12 – 11/12		**Sunita L. Williams**		Kevin A. Ford
				Yuri I. Malenchenko		Oleg V. Novitskiy
				Akihiko Hoshide		Evgeny I. Tarelkin
34		11/12 – 03/13		**Kevin A. Ford**		Thomas H. Marshburn
				Oleg V. Novitskiy		Chris A. Hadfield
				Evgeny I. Tarelkin		Roman Y. Romanenko

Commander in bold

 United States, Russia, Canada, Germany, France, Japan, Belgium, Italy, Netherlands

Table 22 – Shuttle Flights to the International Space Station

Shuttle Flight	Dates	Shuttle Crewmembers			
STS-88 (E)	12/04/98 - 12/15/98	CDR	Robert D. Cabana	MS2	Nancy J. Currie
		PLT	Frederick W. Sturckow	MS3	James H. Newman
		MS1	Jerry L. Ross	MS4	Sergei K. Krikalev
STS-96 (D)	05/27/99 - 06/06/99	CDR	Kent V. Rominger	MS3	Daniel T. Barry
		PLT	Rick D. Husband	MS4	Julie Payette
		MS1	Tamara E. Jernigan	MS5	Valery I. Tokarev
		MS2	Ellen Ochoa		
STS-101 (A)	05/19/00 - 05/29/00	CDR	James D. Halsell, Jr.	MS3	James S. Voss
		PLT	Scott J. Horowitz	MS4	Susan J. Helms
		MS1	Mary E. Weber	MS5	Yury V. Usachev
		MS2	Jeffrey N. Williams		
STS-106 (A)	09/08/00 - 09/20/00	CDR	Terrence W. Wilcutt	MS3	Daniel C. Burbank
		PLT	Scott D. Altman	MS4	Yuri Malenchenko
		MS1	Edward T. Lu	MS5	Boris V. Morokov
		MS2	Richard A. Mastracchio		
STS-92 (D)	10/11/00 - 10/24/00	CDR	Brian Duffy	MS3	Peter J.K. Wisoff
		PLT	Pamela A. Melroy	MS4	Michael E. Lopez-Alegria
		MS1	Leroy Chiao	MS5	Koichi Wakata
		MS2	William S. McArthur		
STS-97 (E)	11/30/00 - 12/11/00	CDR	Brent W. Jett	MS2	Marc Garneau
		PLT	Michael J. Bloomfield	MS3	Carlos I. Noriega
		MS1	Joseph R. Tanner		
STS-98 (A)	02/07/01 - 02/20/01	CDR	Kenneth D. Cockrell	MS2	Marsha S. Ivins
		PLT	Mark L. Polansky	MS3	Thomas D. Jones
		MS1	Robert L. Curbeam		
STS-102 (D)	03/08/01 - 03/21/01	CDR	James D. Wetherbee	MS3	Yury V. Usachev
		PLT	James M. Kelly	MS4	James S. Voss
		MS1	Andrew S. W. Thomas	MS5	Susan J. Helms
		MS2	Paul W. Richards		
STS-100 (E)	04/19/01 - 05/01/01	CDR	Kent V. Rominger	MS3	Scott E. Parazynski
		PLT	Jeffrey S. Ashby	MS4	Umberto Guidoni
		MS1	Chris Hadfield	MS5	Yuri Lonchakov
		MS2	John L. Phillips		
STS-104 (A)	07/12/01 - 07/24/01	CDR	Steven W. Lindsey	MS2	Janet L. Kavandi
		PLT	Charles O. Hobaugh	MS3	James F. Reilly
		MS1	Michael L. Gernhardt		
STS-105 (D)	08/10/01 - 08/22/01	CDR	Scott J. Horowitz	MS3	Frank L. Culbertson, Jr.
		PLT	Frederick W. Sturckow	MS4	Mikhail Turin
		MS1	Daniel T. Barry	MS5	Vladimir N. Dezhurov
		MS2	Patrick G. Forrester		
STS-108 (E)	12/05/01 - 12/17/01	CDR	Dominic L. Gorie	MS3	Yuri I. Onufrienko
		PLT	Mark E. Kelly	MS4	Carl E. Walz
		MS1	Linda M. Godwin	MS5	Daniel W. Bursch
		MS2	Daniel M. Tani		
STS-110 (A)	04/08/02 - 04/19/02	CDR	Michael J. Bloomfield	MS3	Ellen L. Ochoa
		PLT	Stephen N. Frick	MS4	Lee M. E. Morin
		MS1	Jerry L. Ross	MS5	Rex J. Walheim
		MS2	Steven L. Smith		
STS-111 (E)	05/05/02 - 06/19/02	CDR	Kenneth D. Cockrell	MS3	Valery G. Korzun
		PLT	Paul S. Lockhart	MS4	Peggy A. Whitson
		MS1	Franklin Chang-Diaz	MS5	Sergei Y. Treshchev,
		MS2	Philippe R. Perrin		
STS-112 (A)	10/07/02 - 10/18/02	CDR	Jeffrey S. Ashby	MS2	Sandra H. Magnus
		PLT	Pamela A. Melroy	MS3	Piers J. Sellers
		MS1	David A. Wolf	MS4	Fyodor N. Yurchikhin

Shuttle Flight	Dates		Shuttle Crewmembers		
STS-113 (E)	11/23/02 - 12/07/02	CDR PLT MS1 MS2	James D. Wetherbee Paul S. Lockhart Michael E. Lopez-Alegria John B. Herrington	MS3 MS4 MS5	Kenneth D. Bowersox Nikolai M. Budarin Donald R. Pettit
STS-114 (D)	07/26/05 - 08/09/05	CDR PLT MS1 MS2	Eileen M. Collins James M. Kelly Soichi Noguchi Stephen K. Robinson	MS3 MS4 MS5	Andrew S.W. Thomas Wendy B. Lawrence Charles J. Camarda
STS-121 (D)	07/04/06 - 07/17/06	CDR PLT MS1 MS2	Steven W. Lindsey Mark E. Kelly Michael E. Fossum Lisa M. Nowak	MS3 MS4 MS5	Stephanie D. Wilson Piers J. Sellers Thomas A. Reiter
STS-115 (A)	09/09/06 - 09/21/06	CDR PLT MS1	Brent W. Jett, Jr. Christopher J. Ferguson Steven G. MacLean	MS2 MS3 MS4	Daniel C. Burbank Heidemarie Stefanyshyn-Piper Joseph R. Tanner
STS-116 (D)	12/09/06 - 12/22/06	CDR PLT MS1 MS2	Mark L. Polansky William A. Oefelein Nicholas J.M. Patrick Robert L. Curbeam, Jr.	MS3 MS4 MS5	A. Christer Fuglesang Joan E. Higginbotham Sunita Williams
STS-117 (A)	06/08/07 - 06/22/07	CDR PLT MS1 MS2	Frederick W. Sturckow Lee J. Archambault Patrick G. Forrester Steven R. Swanson	MS3 MS4 MS5	John D. Olivas James F. Reilly Clayton C. Anderson
STS-118 (E)	08/08/07 - 08/21/07	CDR PLT MS1 MS2	Scott J. Kelly Charles O. Hobaugh Tracy E. Caldwell Richard A. Mastracchio	MS3 MS4 MS5	Dafydd R. Williams Barbara R. Morgan B. Alvin Drew
STS-120 (D)	10/23/07 - 11/07/07	CDR PLT MS1 MS2	Pamela A. Melroy George D. Zamka Douglas H. Wheelock Stephanie D. Wilson	MS3 MS4 MS5	Scott E. Parazynski Paolo A. Nespoli Daniel M. Tani
STS-122 (A)	02/07/08 - 02/20/08	CDR PLT MS1 MS2	Stephen N. Frick Alan G. Poindexter Leland D. Melvin Rex J. Walheim	MS3 MS4 MS5	Hans W. Schlegel Stanley G. Love Léopold Eyharts
STS-123 (E)	03/11/08 - 03/26/08	CDR PLT MS1 MS2	Dominic L.P. Gorie Gregory H. Johnson Robert L. Behnken Michael J. Foreman	MS3 MS4 MS5	Richard M. Linnehan Takao Doi, JAXA Garrett E. Reisman
STS-124 (D)	05/31/08 - 06/14/08	CDR PLT MS1 MS2	Mark E. Kelly Kenneth T. Ham Karen L. Nyberg Ronald J. Garan, Jr.	MS3 MS4 MS5	Michael E. Fossum Akihiko Hoshide Gregory E. Chamitoff
STS-126 (E)	11/14/08 - 11/30/08	CDR PLT MS1 MS2	Christopher J. Ferguson Eric A. Boe Heidemarie Stefanyshyn-Piper Stephen G. Bowen	MS3 MS4 MS5	Donald R. Pettit Robert S. Kimbrough Sandra H. Magnus
STS-119 (D)	03/15/09 - 03/28/09	CDR PLT MS1 MS2	Lee J. Archambault Dominic A. Antonelli Joseph M. Acaba Steven R. Swanson	MS3 MS4 MS5	Richard R. Arnold John L. Phillips Koichi Wakata
STS-127 (E)	07/15/09 - 07/31/09	CDR PLT MS1 MS2	Mark L. Polansky Douglas G. Hurley Christopher J. Cassidy Julie Payette	MS3 MS4 MS5	Thomas H. Marshburn David A. Wolf Timothy L. Kopra

Shuttle Flight	Dates	Shuttle Crewmembers			
STS-128 (D)	08/28/09 - 09/11/09	CDR PLT MS1 MS2	Frederick W. Sturckow Kevin A. Ford Patrick G. Forrester José M. Hernández	MS3 MS4 MS5	A. Christer Fuglesang John D. Olivas Nicole P. Stott
STS-129 (E)	11/16/09 - 11/27/09	CDR PLT MS1	Charles O. Hobaugh Barry E. Wilmore Leland D. Melvin	MS2 MS3 MS4	Randolph J. Bresnik Michael J. Foreman Robert L. Satcher, Jr.
STS-130 (E)	02/08/10 - 02/21/10	CDR PLT MS1	George D. Zamka Terry W. Virts, Jr. Kathryn P. Hire	MS2 MS3 MS4	Stephen K. Robinson Nicholas J.M. Patrick Robert L. Behnken
STS-131 (D)	04/05/10 - 04/20/10	CDR PLT MS1 MS2	Alan G. Poindexter James P. Dutton, Jr. Richard Mastracchio Dorothy M. Metcalf-Lindenburger	MS3 MS4 MS5	Stephanie D. Wilson Naoko Yamazaki Clayton C. Anderson
STS-132 (A)	05/14/10 - 05/26/10	CDR PLT MS1	Kenneth T. Ham Dominic A. Antonelli Garrett E. Reisman	MS2 MS3 MS4	Michael T. Good Stephen G. Bowen Piers J. Sellers
STS-133 (D)	02/24/11 - 03/09/11	CDR PLT MS1	Steven W. Lindsey Eric A. Boe Nicole P. Stott	MS2 MS3 MS4	B. Alvin Drew Michael R. Barratt Stephen G. Bowen
STS-134 (E)	05/16/11 - 06/01/11	CDR PLT MS1	Mark E. Kelly Gregory H. Johnson E. Michael Fincke	MS2 MS3 MS4	Roberto Vittori Andrew J. Feustel Gregory E. Chamitoff
STS-135 (A)	07/08/11 - 07/21/11	CDR PLT	Christopher J. Ferguson Douglas G. Hurley	MS1 MS2	Sandra H. Magnus Rex J. Walheim

(E) – Endeavour (D) – Discovery (A) – Atlantis **CDR** – Commander **PLT** – Pilot **MS** – Mission Specialist

Table 23 – Soyuz Flights to the International Space Station

Spacecraft/ISS Mission (Taxi Flight)	Launch Date	Land Date
Soyuz TM-31/1S	10/31/00	05/06/01
Soyuz TM-32/2S	02/28/01	10/31/01
Soyuz TM-33/3S (France: Andromede)	10/21/01	05/05/02
Soyuz TM-34/4S (Italy: Marco Polo)	04/25/02	11/10/02
Soyuz TMA-1/5S (Belgium: Odissea)	10/30/02	05/04/03
Soyuz TMA-2/6S	04/26/03	10/28/03
Soyuz TMA-3/7S (Spain: Cervantes)	10/18/03	04/30/04
Soyuz TMA-4/8S (The Netherlands: Delta)	04/19/04	10/24/04
Soyuz TMA-5/9S	10/14/04	04/24/05
Soyuz TMA-6/10S (Italy: Eneide)	04/15/05	10/11/05
Soyuz TMA-7/11S	10/01/05	04/08/06
Soyuz TMA-8/12S	03/30/06	09/28/06
Soyuz TMA-9/13S	09/18/06	04/21/07
Soyuz TMA-10/14S	04/07/07	10/21/07
Soyuz TMA-11/15S	10/10/07	04/19/08
Soyuz TMA-12/16S	04/08/08	12/23/08
Soyuz TMA-13/17S	10/14/08	04/08/09
Soyuz TMA-14/18S	03/26/09	10/11/09
Soyuz TMA-15/19S	05/27/09	12/01/09
Soyuz TMA-16/20S	09/30/09	03/18/10
Soyuz TMA-17/21S	12/20/09	06/01/10
Soyuz TMA-18/22S	04/02/10	09/25/10
Soyuz TMA-19/23S	06/15/10	11/26/10
Soyuz TMA-01M/24S	10/07/10	03/16/11
Soyuz TMA-20/25S	12/15/10	05/24/11
Soyuz TMA-21/26S	04/04/11	09/16/11
Soyuz TMA-02M/27S	06/07/11	11/22/11
Soyuz TMA-22/28S	11/14/11	04/27/12
Soyuz TMA-03M/29S	12/21/11	07/01/12
Soyuz TMA-04M/30S	05/15/12	September 2012
Soyuz TMA-05M/31S	07/15/12	November 2012

Table 24 – International Space Station Progress Supply Missions

Spacecraft/ISS Mission	Dock Date	Undock Date
Progress M1-3/1P	08/08/00	11/01/00
Progress M1-4/2P	11/18/00	02/08/01
Progress M-44/3P	02/28/01	04/16/01
Progress M1-6/4P	05/23/01	08/22/01
Progress M-45/5P	08/23/01	11/22/01
Progress M1-7/6P	11/28/01	03/19/02
Progress M1-8/7P	03/24/02	06/25/02
Progress M-46/8P	06/29/02	09/24/02
Progress M1-9/9P	09/29/02	02/01/03
Progress M-47/10P	02/04/03	08/28/03
Progress M1-10/11P	06/11/03	09/04/03
Progress M-48/12P	08/31/03	01/28/04
Progress M1-11/13P	01/31/04	05/24/04
Progress M-49/14P	05/27/04	07/30/04
Progress M-50/15P	08/14/04	12/22/04
Progress M-51/16P	12/25/04	02/27/05
Progress M-52/17P	03/02/05	06/16/05
Progress M-53/18P	06/19/05	09/07/05
Progress M-54/19P	09/10/05	03/03/06
Progress M-55/20P	12/23/05	06/19/06
Progress M-56/21P	04/26/06	09/19/06
Progress M-57/22P	06/26/06	01/16/07
Progress M-58/23P	10/26/06	03/27/07
Progress M-59/24P	01/20/07	08/01/07
Progress M-60/25P	05/15/07	09/19/07
Progress M-61/26P	08/05/07	12/22/07
Progress M-62/27P	12/26/07	02/04/08
Progress M-63/28P	02/05/08	04/07/08
Progress M-64/29P	05/16/08	09/01/08
Progress M-65/30P	09/17/08	11/14/08
Progress M-01M/31P	11/30/08	02/06/09
Progress M-66/32P	02/13/09	05/06/09
Progress M-02M/33P	05/12/09	06/30/09
Progress M-67/34P	07/29/09	09/21/09
Progress M-03M/35P	10/18/09	04/22/10
Progress M-04M/36P	02/04/10	05/10/10
Progress M-05M/37P	05/01/10	10/25/10
Progress M-06M/38P	07/04/10	08/31/10
Progress M-07M/39P	09/12/10	02/20/11
Progress M-08M/40P	10/30/10	01/24/11
Progress M-09M/41P	01/30/11	04/22/11
Progress M-10M/42P	04/29/11	10/29/11
Progress M-11M/43P	06/23/11	08/23/11
Progress M-13M/45P	11/02/11	01/23/12
Progress M-14M/46P	01/28/12	04/19/12
Progress M-15M/47P	04/22/12	07/30/12

Table 25 – Automated Transfer Vehicle (ATV)

Spacecraft/ISS Mission	Dock Date	Undock Date
Jules Verne/ATV-1	04/03/08	09/05/08
Johannes Kepler/ATV-2	02/24/11	05/20/11
Edoardo Amaldi/ATV-3	03/28/12	September 2012

Table 26 – H-II Transfer Vehicle (HTV)

Spacecraft/ISS Mission	Berthing Date	Unberthing Date
Kounotori 1/HTV-1	09/17/09	10/30/09
Kounotori 2/HTV-2	01/27/11	03/28/11
Kounotori 3/HTV-3	07/27/12	September 2012

www.ingramcontent.com/pod-product-compliance
Lightning Source LLC
Chambersburg PA
CBHW081235180526
45171CB00005B/434